THE CYRUS CYLINDER

'The Cyrus Cylinder is one of the most important records of antiquity, from the greatest of the near eastern empires: the Achaemenid Persian. The Cylinder is fascinating for the story of its discovery, its reconstruction and later history, even its forgery on Chinese bone. This presentation of it by several experts, fully illustrated and in colour, offers a great deal for any reader engaged by how we reconstruct antiquity, as well as for dedicated scholars.'

SIR JOHN BOARDMAN, FBA,
EMERITUS LINCOLN PROFESSOR OF CLASSICAL ART
AND ARCHAEOLOGY, UNIVERSITY OF OXFORD

'*The Cyrus Cylinder* represents a very significant addition to existing studies of this iconic object, which only seems to grow in stature with the passage of time. With reference to Irving Finkel's observations, it is of great interest to learn that the text existed in more than one format. That is to say that two newly identified fragments from a conventional tablet show that Cyrus' exceptional message was not only written on barrel-shaped cylinders that were intended for interment in the foundations of major structures, but that it was also written on large, flat tablets that were very possibly intended for public display. In line with certain statements in the Book of Ezra, this finding could also lend new authority to a supposition that Cyrus issued separate proclamations addressed to separate components of the population of Babylon.'

DAVID STRONACH, OBE,
PROFESSOR EMERITUS OF NEAR EASTERN ARCHAEOLOGY,
UNIVERSITY OF CALIFORNIA, BERKELEY

THE CYRUS CYLINDER

The King of Persia's Proclamation from Ancient Babylon

EDITED BY IRVING FINKEL

I.B. TAURIS
LONDON · NEW YORK

PUBLISHED BY I.B. TAURIS & CO. LTD
IN ASSOCIATION WITH THE IRAN HERITAGE FOUNDATION

The publishers gratefully acknowledge the support of the Iran Heritage Foundation towards the publication of this book.

Published in 2013 by I.B. Tauris & Co. Ltd
6 Salem Road, London W2 4BU
175 Fifth Avenue, New York NY 10010

www.ibtauris.com

Distributed in the United States and Canada exclusively by Palgrave Macmillan
175 Fifth Avenue, New York NY 10010

Copyright editorial selection and Introduction © 2013 Irving Finkel
Copyright individual chapters © 2013 John Curtis, Irving Finkel, Shahrokh Razmjou, St John Simpson, Jonathan Taylor, Jonathan Tubb

The right of Irving Finkel to be identified as the editor of this work has been asserted by him in accordance with the Copyright, Designs and Patents Act 1988.

All rights reserved. Except for brief quotations in a review, this book, or any part thereof, may not be reproduced, stored in or introduced into a retrieval system, or transmitted, in any form or by any means, electronic, mechanical, photocopying, recording or otherwise, without the prior written permission of the publisher.

ISBN 978 1 78076 063 6

A full CIP record for this book is available from the British Library
A full CIP record is available from the Library of Congress

Library of Congress Catalog Card Number: available

Designed and typeset in 11 on 16 Jenson by illuminati, Grosmont
Printed and bound in Spain

Contents

LIST OF ILLUSTRATIONS   vii

Preface   xi
JONATHAN TUBB

Introduction   1
IRVING FINKEL

CHAPTER 1   The Cyrus Cylinder: the Babylonian perspective   4
IRVING FINKEL

CHAPTER 2   The Cyrus Cylinder: discovery   35
JONATHAN TAYLOR

CHAPTER 3   The Cyrus Cylinder: display and replica   69
ST JOHN SIMPSON

CHAPTER 4   The Cyrus Cylinder: the creation of an icon and its loan to Tehran   85
JOHN CURTIS

CHAPTER 5   The Cyrus Cylinder: a Persian perspective   104
SHAHROKH RAZMJOU

Afterword IRVING FINKEL	127
APPENDIX Transliteration of the Cyrus Cylinder text IRVING FINKEL	129
REFERENCES	137
INDEX	141

List of illustrations

1. An early photograph of the Cyrus Cylinder, as it appeared in Hormuzd Rassam's *Asshur and the Land of Nimrod* — 8
2. The Cyrus Cylinder today — 8
3. Map showing the location of Babylon and Pasargadae — 10
4. Hormuzd Rassam, the discoverer of the Cyrus Cylinder — 12
5. Photograph of one end of the Cyrus Cylinder taken in the British Museum conservation laboratory — 13
6. The Cyrus Cylinder photographed shortly after the addition of the Yale fragment — 14
7. The first 'flattened out' photograph of the Cyrus Cylinder — 16
8. A modern version of this same view — 17
9. The recently identified fragments of the Babylonian tablet that duplicates the Cyrus Cylinder — 19
10. The two troublesome fakes: fossilised Chinese horse bones inscribed with signs from Cyrus' original cuneiform text — 27
11. The forger's crib — 32
12. Sir Henry Creswicke Rawlinson — 36
13. Letter from Rassam to Birch, 20 November 1879 — 37
14. Hormuzd Rassam, who led the British Museum's excavations at Babylon in 1879 — 39
15. The dispatch inventory drawn up by Toma in Babylon — 51
16. The receipt inventory compiled by Pinches of the shipment that arrived at the British Museum — 53
17. Detail from *Map of the Ruins at Babylon*, 1859 — 57
18. Letter from Rawlinson to Pinches, dated 18 September 1879 — 61
19. Letter from Rawlinson to Birch, dated 10 October 1879 — 61

20	Theophilus Goldridge Pinches, cuneiform curator at the British Museum	63
21	The Cyrus Cylinder and three other Babylonian cylinders	65
22	Old gallery photograph of the Cyrus Cylinder	70
23	Another view of the Cyrus Cylinder on its original mount	70
24	A nineteenth-century view of the Cyrus Cylinder	70
25	Sketch elevation showing the position of the Cyrus Cylinder in the Persian Room	72
26	Part of a plaster model of an Achaemenid column capital from Susa	73
27	Sketch plan of the Persian Room	74
28	Sketch elevation showing the position of the Cyrus Cylinder in the post-World War II Persian Room	74
29	Plan of the British Museum galleries indicating the position of the Persian Landing, 1961	75
30	Reverend Norman Sharp with Ali Sami at Persepolis	76
31	Cover of the catalogue of the British Museum exhibition *Royal Persia*	77
32	View and sketch plan of the Iranian Room in 1975	78
33	View of the Cyrus Cylinder case display in the 1975 Iranian Room gallery	79
34	View of the Cyrus Cylinder case display in the temporary 1994 Iranian Room gallery	79
35	View of the Cyrus Cylinder case display in the 1995 Iranian Room gallery	79
36	View of the Cyrus Cylinder case display in the 2007 Rahim Irvani Gallery for Ancient Iran	80
37	Replica of the Cyrus Cylinder in its presentation box	81
38	The crest on the top of the box containing the replica	81
39	Prepaid postcard with image of Mohammad Reza Shah Pahlavi featuring the Cyrus Cylinder	86
40	First-day cover with set of stamps issued to commemorate the founding of the Persian Empire	86
41	Stamps featuring the Cyrus Cylinder issued by Iran, Ethiopia and Romania	87
42	Bronze medal commemorating the founding of the Persian Empire	89
43	Iranian proof silver 75 rial coin and silver medal featuring the Cyrus Cylinder issued to commemorate the founding of the Persian Empire	89
44	Miniature sheet of stamps inscribed 'Ajman State' issued to commemorate the founding of the Persian Empire	90
45	Miniature sheet of stamps inscribed 'State of Oman' issued to commemorate the founding of the Persian Empire	91
46	Imperforate stamp inscribed 'Umm al Qiwain' issued to commemorate the founding of the Persian Empire	92

47	Stamp featuring the Cyrus Cylinder issued by Iran to commemorate the 22nd International Red Cross Conference	92
48	Miniature sheet of stamps issued by Iran to commemorate World Expo 2005 in Aichi, Japan	93
49	Photographers massed behind the case with the Cyrus Cylinder, National Museum of Iran	95
50	General view of the exhibition hall with the Cyrus Cylinder in the National Museum of Iran	96
51	The forecourt of the National Museum of Iran with installations for the Cyrus Cylinder exhibition	97
52	British Museum staff with President Ahmadinejad at the opening of the exhibition in Tehran	98
53	Brass plaque commemorating the exhibition and a cast of the Cyrus Cylinder in the porch of the National Museum of Iran	101
54	The Babylonian *Verse Account*, a satirical attack on Nabonidus and his behaviour	106
55	Photograph of the tomb of Cyrus at Pasargadae	113
56	The stela of Nabonidus in the British Museum	119

Preface

It is a pleasure to signal here the publication of this new volume dedicated to a study of the Cyrus Cylinder, one of the most significant objects in the collections of the British Museum.

The five papers published here have grown out of those presented by British Museum curators at the two-day Cyrus Cylinder Workshop that took place in the British Museum on Wednesday 23 June and Thursday 24 June 2010, and brought together an international team of scholars to discuss new findings and talk over old problems.

We are most grateful to the Iran Heritage Foundation for a generous donation towards underwriting that workshop and to Ali Sattaripour for his generous help in raising sponsorship for the publication of the present volume.

The Cyrus Cylinder has recently been displayed in two exhibitions in the British Museum, *Forgotten Empire* and *Babylon: Myth and Reality*, and has also been loaned to the National Museum in Tehran. We are again indebted to the Iran Heritage Foundation for supporting the touring exhibition of the Cyrus Cylinder across America.

JONATHAN TUBB
Keeper, Department of the Middle East, The British Museum
3 December 2012

Introduction

IRVING FINKEL

THE Cyrus Cylinder is one of the world's best-known cuneiform inscriptions and at the same time one of the most famous archaeological objects in the British Museum in London.

In 539 BC Cyrus II, the great king of Achaemenid Iran, conquered the age-old city of Babylon, and in doing so ushered in a new epoch of ancient history. It was truly a moment of astonishing significance. The city was located on the Euphrates river in the centre of what is today Iraq. By taking the capital Babylon Cyrus inherited not only the city, with all its treasures and traditions, but at the same time a great empire that still retained much of the territory and power that had been so effectively acquired and held by the great kings of the outgoing Neo-Babylonian dynasty, especially Nabopolassar, Nebuchadnezzar II and Nabonidus, the last of whom came to be the final native ruler of Babylonia. Cyrus' conquest was evidently accomplished with a fine awareness of military intelligence coupled with a sense of timing and finesse. The Babylonian inscription on his famous clay cylinder, which was buried in a wall at Babylon, claims in language that has become celebrated that no blood was shed, and that the resident population welcomed his advent with celebration and expressions of loyalty. The conditions that both preceded and followed the conquest – inasmuch as we can grasp them today – will be considered in this book, but the Persian conquest of Babylon has passed into history as a rare example of conquest without slaughter and pillage. The regal narrative as preserved in the clay has been studied, translated, quoted and lauded in a myriad of contexts: one feature that has always compounded its importance has been the fact that it was unique.

In fact this so-called uniqueness has now come to require reconsideration. At the very end of 2009 and the start of the following year work among the collections of cuneiform tablets and fragments in the British Museum brought to light two small pieces of what was once a large tablet of conventional oblong shape inscribed with a text identical – in so far as it is preserved – to that of the Cyrus Cylinder. This discovery clarified, as had sometimes been suspected, that the apparent uniqueness of the Cylinder was merely the accident of discovery, and showed that contemporary versions of the edict were not exclusively for burial but were also circulated as part of Persian state politics. This important discovery prompted the British Museum to host a two-day Cyrus Cylinder Workshop to which interested scholars could be invited to assess the importance of the new discovery, and discuss other issues that surfaced at much the same time. Important here were two fossilised bones from China of uncertain date and manufacture, inscribed with part of the Cyrus Cylinder text, whose genuineness was in doubt, and which in the interval since the workshop have been shown to be certain forgeries. The workshop ran from Wednesday 23 June to Thursday 24 June 2010, and was supported with a generous donation from the Iran Heritage Foundation, to whom all thanks are due. The programme of that occasion was as follows:

WEDNESDAY 23 JUNE 2010

10.00 Registration
10.30 Welcome remarks: *Neil MacGregor*
11.00–13.00 The new tablet fragments and discussions: *Chair, I.L. Finkel, with Sh. Razmjou, W.G. Lambert, M.W. Stolper, Wu Yuhong, P. Daneshmand, M.J. Geller, H. Schaudig, P. Michalowski, J.J. Taylor & C.B.F. Walker*
14.00–14.30 'Cyrus and Cambyses', *M.W. Stolper*
14.30–15.00 'History and Geography in the Victory of Cyrus', *H. Schaudig*
15.00–15.30 'The Style of the Cyrus Cylinder', *P. Daneshmand*

16.15–16.45	'Cyrus, Nabonidus and Ashurbanipal', *P. Michalowski*
16.45–17.15	'The Cylinder as Artefact', *J.J. Taylor*
17.15–17.30	'The Cast of the Cylinder', *St J. Simpson*

THURSDAY 24 JUNE 2010

10.30–13.00	The Chinese bones and discussions: *Chair, I.L. Finkel, with Sh. Razmjou, W.G. Lambert, M.W. Stolper, Wu Yuhong, P. Daneshmand, M.J. Geller, H.Schaudig, P. Michalowski, J.J. Taylor & C.B.F. Walker*
14.00–14.30	'On the Chinese Bones', *Wu Yuhong*
14.30–14.45	'The Context of Some Royal Proclamations', *J.E. Reade*
14.45–15.15	'Cyrus in the Old Testament', *H. Williamson*
15.15–15.30	'Cyrus in the Babylonian Talmud', *M.J. Geller*
16.15–16.45	'Cyrus in Islamic Tradition', *T. Daryaee, read by V.S. Curtis*
16.45–17.15	'The Cyrus Cylinder as an Iranian Icon', *J.E. Curtis*
17.15	Summary and closing remarks
18.00–19.00	Public presentation in the BP Lecture Theatre of the findings of the Cyrus Cylinder Workshop, *with contributions by Neil MacGregor, Irving Finkel, Matthew Stolper & John Curtis*

The present book has to some extent grown out of the discussions during that workshop, but the subject matter has been restricted to what might be summed up as 'The Cyrus Cylinder and the British Museum'. All contributions have been written by staff of the Middle East Department, namely John Curtis, Irving Finkel, Jonathan Taylor, Shahrokh Razmjou (now of Tehran University) and St John Simpson. Here, then, are investigated the full story of the original find at Babylon by Hormuzd Rassam – inasmuch as museum records supply details – and the history of its study and publication within the Museum, as well as an up-to-date treatment of the cuneiform inscription itself with the benefit of the newly found material. In addition, the history of the Cylinder's public exhibition within the museum galleries and the historic loan to the museum in Tehran are fully accounted for, as well as some consideration of the significance of Cyrus' famous declaration in the wider world.

I

The Cyrus Cylinder: the Babylonian perspective

IRVING FINKEL

The Cyrus Cylinder in translation

This book begins with a translation into English of the Babylonian text of the Cyrus Cylinder (*figures* 1–2). The translation has been made directly from a study of the original document in the British Museum, and incorporates new words or parts of words that are provided by the two recently discovered duplicating tablet fragments. The result is that the following translation is as full and up to date a rendering of the Cyrus proclamation as can at present be made. As is discussed below, the new evidence from the tablet fragments includes the name of the Babylonian scribe who was responsible for copying out the tablet. The line of text that gives this information is translated at the end.

Translation of the Cyrus Cylinder

1 [When … Mar]duk, king of the whole of heaven and earth, the …… who, in his … , lays waste his ……
2 [……………………………………………………] broad? in intelligence, …… who inspects (?) the wor]ld quarters (regions)
3 [……………………………………………………] his [first]born (=Belshazzar), a low person, was put in charge of his country,
4 but [……………………………………………………] he set [a (…) counter]feit over them.

5   He ma[de] a counterfeit of Esagil, [and ....] ... for Ur and the rest of the cult-cities.

6   Rites inappropriate to them, [impure] fo[od-offerings ....] disrespectful [...] were daily gabbled, and, as an insult,

7   he brought the daily offerings to a halt; he inter[fered with the rites and] instituted [......] within the sanctuaries. In his mind, reverential fear of Marduk, king of the gods, came to an end.

8   He did yet more evil to his city every day; ... his [people ....], he brought ruin on them all by a yoke without relief.

9   Enlil-of-the-gods became extremely angry at their complaints, and [...] their territory. The gods who lived within them left their shrines,

10 angry that he had made (them) enter into Shuanna (Babylon). Ex[alted Marduk, Enlil-of-the-Go]ds, relented. He changed his mind about all the settlements whose sanctuaries were in ruins,

11 and the population of the land of Sumer and Akkad who had become like corpses, and took pity on them. He inspected and checked all the countries,

12 seeking for the upright king of his choice. He took the hand of Cyrus, king of the city of Anshan, and called him by his name, proclaiming him aloud for the kingship over all of everything.

13 He made the land of Guti and all the Median troops prostrate themselves at his feet, while he shepherded in justice and righteousness the black-headed people

14 whom he had put under his care. Marduk, the great lord, who nurtures his people, saw with pleasure his fine deeds and true heart,

15 and ordered that he should go to Babylon. He had him take the road to Tintir (Babylon), and, like a friend and companion, he walked at his side.

16 His vast troops whose number, like the water in a river, could not be counted, were marching fully armed at his side.

17 He had him enter without fighting or battle right into Shuanna; he saved his city Babylon from hardship. He handed over to him Nabonidus, the king who did not fear him.

18 All the people of Tintir, of all Sumer and Akkad, nobles and governors, bowed down before him and kissed his feet, rejoicing over his kingship and their faces shone.

19 The lord through whose help all were rescued from death and who saved them all from distress and hardship, they blessed him sweetly and praised his name.

20 I am Cyrus, king of the universe, the great king, the powerful king, king of Babylon, king of Sumer and Akkad, king of the four quarters of the world,

21 son of Cambyses, the great king, king of the city of Anshan, grandson of Cyrus, the great king, ki[ng of the ci]ty of Anshan, descendant of Teispes, the great king, king of the city of Anshan,

22 the perpetual seed of kingship, whose reign Bel (Marduk) and Nabu love, and with whose kingship, to their joy, they concern themselves. When I went as harbinger of peace i[nt]o Babylon

23 I founded my sovereign residence within the palace amid celebration and rejoicing. Marduk, the great lord, bestowed on me as my destiny the great magnanimity of one who loves Babylon, and I every day sought him out in awe.

24 My vast troops were marching peaceably in Babylon, and the whole of [Sumer] and Akkad had nothing to fear.

25 I sought the safety of the city of Babylon and all its sanctuaries. As for the population of Babylon [..., w]ho as if without div[ine intention] had endured a yoke not decreed for them,

26 I soothed their weariness; I freed them from their bonds(?). Marduk, the great lord, rejoiced at [my good] deeds,

27 and he pronounced a sweet blessing over me, Cyrus, the king who fears him, and over Cambyses, the son [my] issue, [and over] my all my troops,

28 that we might live happily in his presence, in well-being. At his exalted command, all kings who sit on thrones,

29 from every quarter, from the Upper Sea to the Lower Sea, those who inhabit [remote distric]ts (and) the kings of the land of Amurru who live in tents, all of them,

30 brought their weighty tribute into Shuanna, and kissed my feet. From [Shuanna] I sent back to their places to the city of Ashur and Susa,

31 Akkad, the land of Eshnunna, the city of Zamban, the city of Meturnu, Der, as far as the border of the land of Guti – the sanctuaries across the river Tigris – whose shrines had earlier become dilapidated,

32 the gods who lived therein, and made permanent sanctuaries for them. I collected together all of their people and returned them to their settlements,

33 and the gods of the land of Sumer and Akkad which Nabonidus – to the fury of the lord of the gods – had brought into Shuanna, at the command of Marduk, the great lord,

34 I returned them unharmed to their cells, in the sanctuaries that make them happy. May all the gods that I returned to their sanctuaries,

35 every day before Bel and Nabu, ask for a long life for me, and mention my good deeds, and say to Marduk, my lord, this: 'Cyrus, the king who fears you, and Cambyses his son,

36 may they be the provisioners of our shrines until distant (?) days, and the population of Babylon call blessings on my kingship. I have enabled all the lands to live in peace.'

37 Every day I increased by [... ge]ese, two ducks and ten pigeons the [former offerings] of geese, ducks and pigeons.

38 I strove to strengthen the defences of the wall Imgur-Enlil, the great wall of Babylon,

39 and [I completed] the quay of baked brick on the bank of the moat which an earlier king had bu[ilt but not com]pleted its work.

40 [I ...... which did not surround the city] outside, which no earlier king had built, his workforce, the levee [from his land, in/int]o Shuanna.

41 [.. with bitum]en and baked brick I built anew, and [completed] its [work].

42 [...] great [doors of cedar wood] with bronze cladding,

43 [and I installed] all their doors, threshold slabs and door fittings with copper parts. [.......................]. I saw within it an inscription of Ashurbanipal, a king who preceded me;

44 [...] in its place. May Marduk, the great lord, present to me as a gift a long life and the fullness of age,

45 [a secure throne and an enduring rei]gn, [and may I ...... in] your heart forever.

The scribal note from the tablet:
[Written and check]ed [from a...]; (this) tablet (is) of Qīšti-Marduk, son of [...].

CLAY CYLINDER OF CYRUS CONTAINING AN ACCOUNT OF THE CAPTURE OF THE CITY OF BABYLON. 538 B. C.

1 (*top*) An early photograph of the Cyrus Cylinder, as it appeared in Hormuzd Rassam's book *Asshur and the Land of Nimrod* (1897), plate opposite p. 268.

2 (*bottom*) The Cyrus Cylinder today.

TABLE I.I Names of gods, people and places mentioned in the Cyrus Cylinder inscription, in order of appearance

Marduk	The city god of Babylon and the patron of the Neo-Babylonian kings of Babylon
Belshazzar	The firstborn son of Nabonidus, the last native king of Babylon (556–539 BC), who acted as regent for his father during his absence in Teima in Arabia
Esagil	The great temple dedicated to Marduk in Babylon, south of the ziggurat tower complex Etemenanki
Enlil	The second most powerful of the ancient gods of Mesopotamia, whose place was usurped by Marduk
Shuanna	A name for the city of Babylon, here standing for the whole city but in fact that name of the southernmost quarter
Sumer and Akkad	The old names for southern Mesopotamia, later Babylonia
Tintir	The old Sumerian name for the city of Babylon
Anshan	Ancient Elamite city, modern Tal-e Malyan, northwest of Shiraz in southern Iran
Guti	The name for the inhabitants of the area between the Zagros Mountains and the River Tigris, referring specifically to Iranians, including the Medes and their companions
Nabu	The god of writing, son of Marduk and his wife Zarpanitu
Amurru	The west land
Ashur	One of the old Assyrian capital cities in the north of Iraq, on the upper reaches of the River Tigris
Susa	Elamite capital city in southwest Iran
Eshnunna	Central Babylonian city on the Diyala river in Iraq
Zamban	A city in the northeast
Meturnu	A city located roughly between Zamban and Eshnunna
Der	An ancient city located east of the Tigris river on the border between Sumer and Elam
Tigris	The river that with the Euphrates defined ancient Mesopotamia, the 'land between the rivers'
Imgur-Enlil	The famous Inner Wall of the city of Babylon
Ashurbanipal	The last great king of Assyria (685–627 BC), warrior-librarian, who undertook restoration work in Babylon and left an account of it buried for the future that Cyrus later discovered

The localities (with the exception of Susa) are mentioned in connection with the restoration of temples, and were in eastern and northern Mesopotamia, in territories formerly ruled over by Nabonidus (*table 1.1*; *figure 3*).

3 Map showing the location of Babylon and Pasargadae.
M. Chudasama.

The Cyrus Cylinder as object

The Cyrus Cylinder is not, in fact, a cylinder at all. Rather is it barrel-shaped, slightly swollen in the middle, and it is made of pale-coloured clay that was almost certainly fired in antiquity to ensure its long-term survival. Neither the cylinder nor its inscription was meant for human eyes, for the object was a record that was buried invisibly in a major wall of the city of Babylon. There it could be read by the gods, and, equally importantly, by any later king who might engage in building, or rebuilding, at the same spot, and who might come upon Cyrus' statement. As Jonathan Taylor shows below (Chapter 2), in secreting such a cylinder to commemorate important religious building Cyrus was following closely in the footsteps of the oldest kings of Mesopotamia. That Cyrus should follow Babylonian precedent in this way was certainly no accident, but must be seen as part of a much wider policy of doing everything to demonstrate that, while a Persian from beyond the eastern border, he knew how to behave like a Babylonian in matters of religion, administration and tradition in general. For these reasons it must have been a source of satisfaction to him, or at least to his officials, that work on restoring the tumbledown foundations uncovered a similar inscription of Ashurbanipal, the great Assyrian ruler who restored the earlier destruction brought about in the sacred city by his own ancestors, and left an account of it to be discovered by the Persians.

The Cylinder is inscribed with forty-five lines of Babylonian cuneiform script. Cuneiform is the oldest known writing system in the world and it had been in use within the Mesopotamian heartland from the end of the fourth millennium BC (Walker 1987). In its developed form cuneiform was used to write the Sumerian and Babylonian languages with equal facility, as well as a handful of other contemporary languages in the Middle Eastern milieu, including Elamite in a somewhat peculiar style, within Cyrus' own homeland. The language in which it is written is ancient Babylonian, a Semitic language related to modern Semitic languages such as Hebrew, Arabic and Aramaic.

The Cylinder is made of two kinds of clay: a slip of finer material wrapped round an inner core of cruder nature, as may be seen in a laboratory photograph (*figure 5*). Under conditions of stress or damage the outer 'veneer' has been apt to become detached from this inner core. Unlike many of the foundation cylinders of the Neo-Babylonian kings the inscription is written out in a contemporary Late Babylonian script. The use of archaising writing – that is, choosing 'old-fashioned' sign forms that had been in use one thousand years earlier – was a deliberate technique often used to underpin the stability and longevity of the royal dynasts: some of King Nebuchadnezzar's royal inscriptions could have been read at sight by the venerable King Hammurabi, who had then been dead for a millennium. Cyrus, however, despite emulating his predecessors' general example, did not opt to follow this path.

4 Hormuzd Rassam (1826–1910), the discoverer of the Cyrus Cylinder.

The cuneiform signs in the Cylinder text are slightly idiosyncratic in shape and proportion. The signs in the Cylinder have a slightly 'squat' look and give the impression of a scribe whose own handwriting came to maturity with less discipline from his teacher than normal. Each line of Babylonian signs is set between and placed clear of rulings which run right across the Cylinder. Examination of the whole text reveals surprising inconsistency in the cuneiform signs, in terms of both size and spacing. To mark the beginning and end of the inscription, which are separated by a short gap, the signs in the first two lines and the last two lines (at least judging by what survives) are larger and more generously spaced out than in the intervening lines 3–43. This in itself is unremarkable. But among the fully preserved lines there are cases of as few as 33 signs per line (line 13) and as many as 65 (line 35); in the latter case the line ends in a mass of cramped-together, small signs that spill out over the rim. This is in contrast with the usually very finished cylinders of the Neo-Babylonian dynasty that Cyrus was clearly emulating. Viewed simply as a scribal achievement, the Cyrus Cylinder is a relatively poor piece of work.

5 Photograph of one end of the Cyrus Cylinder taken in the British Museum conservation laboratory, showing the internal core of clay and the outer layer of clay wrapped around it that characterise its construction.

The joining fragment from Yale

In 1971, the Assyriologist P.-R. Berger made the astonishing discovery that a large fragment of cuneiform cylinder in the Yale Babylonian Collection at Yale University in the United States, and which had long before been published, was part of the British Museum Cyrus Cylinder, and, what is more, actually joined it (Berger 1975: 192; Walker 1972). This Yale fragment had been acquired many years before by Rev. Dr. James B. Nies, a private antiquities collector of Brooklyn, New York, who later presented it together with the rest of his cuneiform collection to Yale University. This generosity led to the publication of a sequence of (to date) ten scholarly publications, while the Cylinder fragment in particular was published in the second volume (*Historical, Religious and Economic Texts and Antiquities*) by James B. Nies himself and Clarence E. Keiser in 1920. In his explanation of the source of these tablets the editor, Albert T. Clay, wrote as follows:

> They were collected by Doctor Nies during the past fifteen years [that is, literally, between 1905 and 1920]. Some were secured by him while in Bagdad; others were purchased in Paris, London, New York and elsewhere. When such objects, of unquestionable value, have found their way into the hands of dealers it seems highly advisable to rescue them, if possible, for science by purchasing them, even though we know that some are the results of illicit excavations by Arabs, and that others may have been purloined from legitimate excavations by workmen. (Nies and Keiser 1920: Editorial Note)

6 The Cyrus Cylinder photographed shortly after the addition of the Yale fragment.

Following this discovery, and its announcement, the Yale fragment was deposited on long-term loan in the British Museum.

As is discussed in Chapter 2, whether or not the Cyrus Cylinder was undamaged on original discovery, the object that reached the British Museum in London was not only incomplete but in pieces, which had to be glued back together by the Museum craftsman (*figure 6*). Examination of the composite 'flat' photograph (*figures 7–8*) – which includes the Yale fragment – shows conveniently how much of the text in fact survives and the shape and size of the portions that are lost.

If, as seems not unlikely, the cylinder was broken at some point after first discovery, fragments from the find might easily have disappeared, and this is no doubt how the piece which was ultimately acquired by Dr Nies came onto the market. We know nothing of the history of the Yale fragment between its discovery in 1879 and about 1905. If the other pieces went first to a dealer in Baghdad, as seems likely, some might well have been used to 'beef up' consignments of tablets for acquisitive buyers with an alluring fragment of a cylinder, and it is always possible that a fragment or fragments of the Cyrus Cylinder might still come to light in an unknown collection somewhere.

Following Professor Berger's discovery that the Yale fragment actually joined the Cyrus Cylinder in London, it was arranged between the (then) keeper of the (then) Department of Western Asiatic Antiquities and the (then) keeper of the Babylonian Collection in the Sterling Memorial Library at Yale University that the Yale fragment should come on long-term loan to the British Museum in exchange for a loan in return of a Babylonian copy of the First Tablet of the Babylonian Creation Epic (BM 93015).

The new fragments

Since its discovery, and increasingly since the time of its publication, the Cyrus Cylinder has been celebrated for its uniqueness. The historical setting in which it came into being and the various messages conveyed within the text itself gradually afforded it status far beyond the normal parameters of ancient inscriptions and archaeology. As attention focused on the Cylinder in different quarters it came to achieve worldwide fame and, ultimately, as discussed by John Curtis below, an iconic quality. Writers on all levels hurried to stress the fact that it stood alone, and that no duplicate text from the hand of Cyrus had ever been discovered.

The point has more than a casual validity; many cuneiform inscriptions, especially those of a historical nature that emanate from the offices of state, have come down to us in multiple copies. This is true to the point that a given document which is damaged or fragmentary can very often be restored from a parallel version that happens to be in better condition at the crucial point. Such historical inscriptions were characteristically composed and written out to convey messages on a widespread scale; building bricks, palace sculptures, public monuments and other suitable surfaces could carry royal inscriptions in cuneiform that proclaimed to the world the name of the building, the god to whom it was dedicated, and the king who had built it. Such inscriptions functioned in an environment where usually only a minority of the population was literate, but their message and implication were nevertheless clear: this was official work, the pride of the nation, carried out with the blessing, if not actual support, of the gods themselves.

Such an idea even governed state messages that were hidden or buried. As Jon Taylor discusses, it had been a Mesopotamian tradition for millennia

7 (*overleaf, left*)
The first 'flattened out' photograph of the Cyrus Cylinder, made in the British Museum in the 1970s after the addition of the Yale fragment.

8 (*overleaf, right*)
A modern version of this same view. This shows the reader conveniently how much of the text is still missing.

Photo: M. Arksey.

that new (or restored) buildings were consecrated with foundation deposits in which records of who built it and when were laid down not only for the god, but for future rulers who might find the building dilapidated, learn who had built it, and undertake its respectful restoration.

The Cyrus Cylinder was just such a text, written out and buried or immolated with appropriate ceremony at a strategic point in the rebuilding of Babylon's major architectural features. The cylinder explained at some length how it was that Cyrus was now and forever King of Babylon and confirmed to one and all that the rebuilding was his work, carried out under the supportive eye of Marduk, King of the Gods. In carrying out such extensive reconstruction and producing this commemorative foundation cylinder Cyrus was following faithfully in the footsteps of his predecessors, not only of the Neo-Babylonian dynasty that he had just displaced, but the Assyrians before them, and indeed the Sumerian rulers of the third millennium BC, who had their own devices for adding a signature to a building and preserving their name and achievements.

That the Cyrus Cylinder was a more or less standard foundation inscription for burial, however, has always been at odds with the narrative which is written on it. For the text embodies content with far wider application than befits an unreachable and unreadable foundation deposit, and it had been proposed by more than one scholar that this particular document can hardly have been a one-off; its multi-purpose message is so well tailored to a Babylonian readership that it must have seen a wider distribution than one invisible cylinder. For this reason the recent discovery within the British Museum cuneiform collections of an ancient duplicate of the Cyrus Cylinder text has far-reaching implications, and has meant a new upswing in interest in the text of the cylinder itself and a reappraisal of its significance. The stature of the discovery is not at all reflected in the physical dimensions of the new material: the two identified fragments are small.

The first duplicating fragment was identified in the British Museum by the late Professor W.G. Lambert on 23 December 2009 and the second by the present writer on 4 January 2010. Both fragments belong in a specific collection of mostly small pieces of cuneiform tablet that arrived in the British Museum in 1881, and registered and numbered within a sub-collection of the Babylon tablets known as the 1881,0830

9 The recently identified fragments of the Babylonian tablet that duplicates the Cyrus Cylinder, the one-sided BM 47134 and the two-sided BM 47176, which gives part of the missing beginning and end of the inscription.

collection (*figure 9*). Later the two fragments were given 'long numbers', so that today they are stored as BM 47134 (1881,0830.656) and BM 47176 (1881,0830.698). The 1881,0830 collection contains about 115 tablets and fragments and is of very mixed content and archaeological provenance. It includes royal cylinders, lexical and grammatical texts, omens and hemerologies, magical and medical texts, mathematical texts, god lists and astrological works, field plans, contracts and accounts, and, finally, school texts. In range it thus constitutes a fairly representative sampling of the greater mass of tablets from southern Iraq that came to light during the excavations of the nineteenth century. This is a summary of the background to the 1881,0830 collection:

> *81-8-30* (mainly BM 46535-47310). Five cases sent from Baghdad on 15 June (Trustees' Minutes, 23 July 1881). We have Daud Thoma's inventory of four cases which he sent to Baghdad on 4 June, for shipment on 15 June (1882 P 3013, cases 1–4). They contained, beside uninscribed objects, tablets

THE BABYLONIAN PERSPECTIVE 19

both from Dailem [ancient Dilbat, about 30 km due south of Babylon] and, implicitly, from the other two sites where he was working, Babylon and Ibrahim al-Khalil [adjoining Birs Nimrud, ancient Borsippa, about 25 km south west of Babylon]. A receipt inventory (1882 P 207), submitted to the Trustees four months later, lists material as from Birs Nimrud and Dailem, without reference to Babylon; it also gives a total of 800 or 900 items, whereas there were over 1400 dispatched, suggesting that one case had been mislaid. Moreover a notice of receipt of the next consignment, 81-11-3 (1881 P 5239 and Trustees' Minutes, 10 December 1881), mentions one tablet which appears identical with 81-8-30, 9 (BM 46543), probably no. 186 in the 81-8-30 receipt inventory, which should come from the Borsippa Nabu Temple (*LIH*, no. 50). With this degree of confusion developing, and cases containing the 81-8-30, 81-11-3, 82-3-23, and 82-5-22 consignments accumulating unnumbered at the BM (1882 P 3013), the practice of making detailed receipt inventories was dropped. There therefore seems no point in pursuing the identity of the fifth box in the 81-8-30 consignment: perhaps it contained bricks left with Plowden by Rassam (1881 P 2286), or was a small one that apparently came from Abu Habbah [ancient Sippar] to Baghdad in June (1882 P 1259), or resulted from repacking in Baghdad. Our best provenance information probably comes from the marks on the tablets, which are mostly unbaked: several are marked D [i.e. Dailem], and the registrar relied on this, although missing at least one, 81-8-30, 622 (BM47101); no tablets marked B. N. [i.e. Birs Nimrud] have been noticed, and the abbreviation I. H. [i.e. Ibrahim al-Khalil] was probably no longer in use. (Reade 1986: xxxi–xxxii)

From this rather patchy evidence the late-nineteenth-century find spot of the two 1881,0830 pieces cannot now be established. While one might naturally think of Babylon itself, the ancient cities of Dilbat or Borsippa are equally possible.

The two fragments under discussion come from one large cuneiform tablet that once carried the same text as was originally inscribed on the Cyrus Cylinder. Despite their modest size, it is quite certain that the fragments come from a flat and conventional tablet, and not a second example of a cylinder. Thus they also establish at the outset that the proclamation existed in more than one format, and we can be confident that the tablet version, part of the state operations of the Achaemenid kingship, was never buried in a wall or in the ground as a message for the future.

From the existence of this newly discovered inscription we are entitled to draw certain conclusions. One is that the subject of the proclamation is likely to have been promulgated far and wide, in the first instance within the former kingdom of Babylonia, but also with some degree of probability throughout the territories of the Achaemenid Empire. Even a cursory glance at the text of the Cylinder shows that the inscription falls into discrete sections, and it is not hard to imagine that there might have been a core account of the conquest of Babylon and the takeover of power which could have extra passages added or adapted to local interest in the creation of other such accounts. This example, written for burial in the foundations at Babylon, devotes a long passage to describing the specific building programme that it commemorates, and might not have been so appropriate in other contexts, both in and out of Mesopotamia proper. In addition, given the use of languages under the Achaemenid umbrella, it is far from unlikely that such official promulgations were also put out in Persian, Aramaic and even Elamite versions, adjusted in detail as was appropriate. It was necessary for everyone to understand that, as it were, history had moved on, that the conquest of mighty Babylon had been effected without bloodshed or rebellion, and that the very gods of Babylon smiled down on the new conqueror and his son to be king after him.

A second, quite unexpected, boon resulting from this discovery is that both fragments, despite their modest size, add something new to the incomplete text known from the Cylinder. The one fragment is one-sided but adds certain crucial information about the Achaemenid view of their religious responsibilities, while the two-sided second fragment comes from the top edge of the tablet and has therefore new material from the very beginning and very end of the inscription, which as luck would have it are the most serious areas of loss in the Cylinder itself.

The fragments are written in an excellent and accomplished professional hand which is quite different from the Cyrus Cylinder script alluded to above. As with the Cylinder, every line is ruled, but here the last line is finished off with a double ruling. The two fragments represent a mere fraction of the original tablet, and their size and nature suggest that the original document might have been broken into very many such pieces. It was a brilliant coup by Wilfred Lambert to identify the first piece, since while the phraseology of the few preserved Babylonian words was obviously literary and concerned with the god Marduk, there are abundant other religious compositions in which such phrases occur. This first identification led directly to the second, since the scribal hand is readily identifiable and there were several key words to work with in that fragment, most especially the name of Ashurbanipal, which is an important feature near the end of the cylinder text, while that Assyrian royal name does not otherwise occur very frequently in Babylonian script. In addition, it should be pointed out that the rich collections of Babylonian source material in the British Museum from the nineteenth century include very copious numbers of small pieces where the correct identification as to textual genre is sometimes extremely difficult to establish until fragments can be joined to one another, or to an already identified text. This means that there are many fragments which can be seen to be, roughly speaking, 'administrative' or 'religious' or 'literary,' but very many more where even such a general categorisation cannot be allotted with certainty. It is not unlikely that further fragments of the broken 1881,0830 tablet remain to be identified among the cuneiform material that came from Babylon in the nineteenth century, but although the present writer has made great efforts to identify further pieces, this has been so far without success. While even a handful of additional signs from Cyrus' text could be of great significance, we must be grateful for what we do have.

A third, and unexpected, feature raised by the existence of the new manuscript is that the scribe who wrote it out included his name at the very end, in the one-line colophon (or scribal note) to the text. This can be translated as follows:

[Written and check]ed [from a …]; a tablet of Qīšti-Marduk, son of […]

This scrap of colophon is thus rather informative. The scribe who actually wrote the tablet represented by BM 47134 and BM 47176 was called Qīšti-Marduk, and when complete the line also recorded the name of his father, who was almost certainly also a scribe himself. Qīšti-Marduk is a standard Babylonian name, meaning 'Gift of Marduk', and is sufficiently common that this individual cannot be identified for certain, especially without the name of his father. (It is also possible that the name was rather Iqish-Marduk, since the Babylonian reading of the first two signs is ambiguous.) As already indicated, the tablet is of fine quality, with excellent script and good-quality clay. On the basis of what survives I would classify Qīšti-Marduk's document as an official copy rather than a production that originated in a scribal school or in a private library. Far more probable is that it is a high-standard production stemming from a chancery or office where multiple copies of official documents in diverse formats were produced.

The existence of these new fragments allows other observations to be made about the Cyrus Cylinder which were scarcely feasible before. In the first place we can say that, in the writing process that led up to the finished Cyrus Proclamation, tablets preceded cylinders; no one ever composed a cuneiform text on a curvaceous cylinder. One can further assume that there were multiple such cylinders prepared for burial at suitable points in the city's reconstruction. In each case the cylinder inscription would be copied from a 'flat' master copy (and not from memory). This tablet–cylinder relationship probably goes a long way to explain the uneven quality of the Cyrus Cylinder inscription, referred to above. The new tablet pieces are too fragmentary to allow certainty, but it is probable that there were some forty-five lines per side, so that one line of the cylinder would correspond to two – or sometimes three – lines on the tablet exemplar. If the scribe was transposing from a tablet version and, at the same time, spreading the tighter format of that tablet out onto a roomier cylinder support, such inconsistency could be a natural result.

It has long been evident that the content of the Cyrus Cylinder divides naturally into three distinct literary sections, especially given the highly significant shift from third person to first person beginning with line 20. The cylinder text may be considered as follows:

SECTION I, LINES 1–19
- Praise of Marduk; probable creation of kingship and Babylon with Marduk as king of the gods.
- Disastrous reign of Nabonidus; his mad worship of an embodiment of the moon; sacrilegious construction of a fake Esagil; imposition of a fake king in the form of his son Belshazzar; Nabonidus' abandonment of the country; utter disaster for gods and men alike.
- The summons of Cyrus by Marduk to put things aright.

SECTION 2, LINES 20–36
- The Cyrus proclamation: 'I am Cyrus…'
- The bloodless conquest of Babylon.
- Universal approbation.

SECTION 3, LINES 37–45
- Rebuilding of Babylon and formulas to protect the inscription.

One can probably assume that these three elements have separate compositional origins. The first might well derive from a court chronicle or other appraisal of Cyrus' early reign; the second represents the words of Cyrus himself and perhaps ultimately reflects a Persian original, while the third is a straightforward account that derives at least in some measure from the earlier text of Assurbanipal found in the digging.

Thinking beyond this, therefore, if we posit (a) the tablet edition preceding the cylinder version, and (b) separate passages preceding the tablet version, the result is the following schema:

1. Marduk and Babylon 2. 'I am Cyrus…' 3. Cyrus rebuilds Babylon	Texts woven together to form a 90-line inscription on both sides of a single-column tablet	Text transferred to a barrel cylinder to form a version of 45 lines

The text that results is highly appropriate for Babylonian sensitivities. The process of recent history is attributed to the intervention of Marduk, who was certainly not the god worshipped by Cyrus himself. It is, therefore, a question worthy of consideration as to how the text came into being. Since the concluding passages of the Cylinder inscription concern Cyrus' building operations at Babylon, and the Cylinder itself was buried within one of those very structures, the text as we have it cannot have come into existence directly after the Persian conquest of the city but only after a certain interval of time. The principal function of the text is to make the incoming Cyrus the Persian acceptable as king of Babylon to the Babylonian population. This ambition was close to Cyrus' heart, and one could imagine that he personally would see every benefit in a proclamation that would achieve the desired end. From this point of view the Cylinder's authorship makes ready sense as an item of state propaganda: the whole episode has come about under Babylonian divine control; Cyrus and his successors will maintain the status quo and support the cult and temple as they should.

Discussions at the British Museum workshop debated whether the Cyrus Cylinder should be classified as a royal inscription or a building inscription. Some participants argued that it was composed by 'a Babylonian – in Babylonian – in Babylonia – for Babylonians', while others saw it as a Persian production that directly reflects Cyrus' interest and is a masterpiece of political and religious propaganda.

We can be sure that the Persian administration kept itself well informed about what was going on in Babylonia prior to the military takeover in 539 BC. The religious conflict that had led to lasting opposition between Marduk and Sin as state god undoubtedly created deep-seated social and religious tension and a situation that was ripe for exploitation. Persian foreign policy came down firmly on the side of Marduk, and Cyrus' intelligence officers – perhaps with the advice of well-placed Babylonian intellectuals – must have been very sure of the state of play in the country and of the views of those who could be counted on post-invasion.

Cyrus' own words, 'I am Cyrus...', in the second section have had special resonance in conjunction with the celebrated passages in the Old Testament in which Cyrus is seen to have a divine mission to deliver Israel from the Babylonian Exile and enable the Temple in Jerusalem to be rebuilt (Isaiah II 40–48; II Chronicles 36:22–23; Ezra 1 and 6:1–5; see Smith 1963; van der Spek forthcoming). The Book of Ezra famously gives what purports to be the proclamation of Cyrus both in Hebrew (Ezra I:2–4) and Aramaic (Ezra 6:2–5), while a fuller version appears in Josephus' *Antiquities of the Jews* Book XI, chapter 1. The authenticity of these traditions has often been discussed and considered, but the possibility remains that the biblical traditions and the 'I am Cyrus...' passage in the Cyrus Cylinder both reflect official state proclamations that appeared from the Persian administration in many forms soon after arrival. Cyrus' advent might have been heralded by trumpets and banners, with versions in Aramaic as well as Babylonian in circulation. It must be stressed that the Cyrus Cylinder makes no reference to the Jewish Exiles; the restoration of normality within Babylonia involves gods and peoples within the Mesopotamian heartland. Nevertheless, the Old Testament view of the divine use of Cyrus as saviour more than echoes the Persian declaration that Cyrus was serving Marduk in putting things to rights in Babylonia.

An important contribution was that of Harmatta (1971), who was the first to argue that the text of the Cylinder was modelled in appreciable measure on a text of the Assyrian king Assurbanipal, bolstered in this view by the fact that Cyrus refers to finding an inscription of that earlier king during the course of restoration work on the wall Imgur-Enlil (Harmatta 1971; further ideas in Kuhrt 1983, and Michalowski forthcoming). Other scholars have discussed the relationship between the Cyrus Cylinder text and Mesopotamian royal ideas and inscriptions (see especially van der Spek, forthcoming).

On the forged Chinese 'bone texts'

Two fossilised horse bones, each inscribed along the shaft with crude but unmistakable cuneiform signs, came to official attention in China in 1983, and are now in the Palace Museum, Beijing (*figure* 10). After a period of uncertainty as to their interpretation it was established by the late Professor O.R. Gurney in Oxford that one of the inscriptions was a partial duplicate to lines 18–21 of the Cyrus Cylinder. The status of these bones has remained problematic ever since, with a majority of scholars condemning them as forgeries. The issue was discussed at length during the 2010 workshop, and, although arguments were put forward at that time by the present writer to defend their being something more significant than out-and-out fakes, most of the participants dismissed the bones as the work of a modern forger. That both are certainly forgeries was only established in the British Museum on Friday 30 November 2012, four days before the present volume was sent to press. The following pages retain a summary of the earlier arguments for and against taking the bones as some kind of textual witness, since the process by which this conclusion was eventually reached is instructive in the general battle against archaeological forgery.

The bones come to light

Considerable details of the circumstances under which the bones came to light are available in Chinese and English. Attention was first drawn to

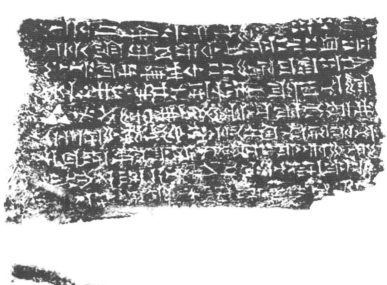

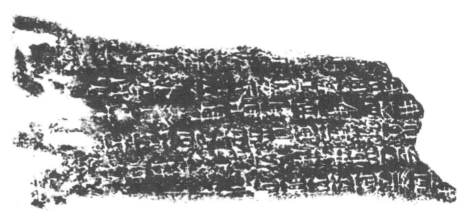

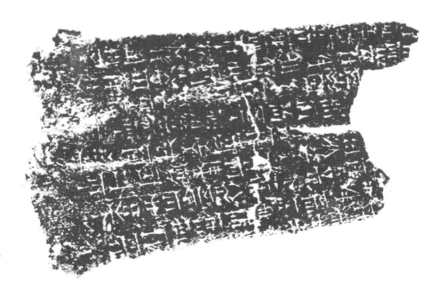

10 The two troublesome fakes: fossilised Chinese horse bones inscribed with signs from Cyrus' original cuneiform text, brought to the attention of Assyriologists by Wu Yuhong (1986); (*top to bottom*) Bone 1; Bone 2, front; Bone 2, back.

them by Xue Shenwei, a traditional doctor, who died in his eighties in 1985. The following summarises the content of the various published accounts that have been available. (New translations of the articles in Chinese were most kindly prepared by Wai Ka Yu, formerly a research fellow in the British Museum Asia Department.)

TABLE 1.2   History and publication of the bones from China

pre-1928	Professor Luo Xuetang was shown a photograph of a rubbing of one of the bones by the collector/dealer Zhang Yi'an in the Peking market. He realised that this must be pre-Shang dynasty writing and tried unsuccessfully to get hold of the original bone.
1928	Doctor Xue Shenwei met Professor Luo Xuetang and learned of the rubbing from him.
1935	Xue Shenwei was attending professionally the wife of one Zhang Yueyan in an unknown part of China. It is thought by Yang Zhi that this individual must be the same as the collector/dealer Zhang Yi'an earlier met by Professor Luo Xuetang. Zhang had a framed rubbing of one of the bones. Xue Shenwei later acquired the original bone from Wang Dongting, an antique dealer. Xue Shenwei showed it to Luo Xuetang, who told him that there were originally two bones, the other being with Ke Yanling, a former student of Professor Luo Xuetang. Xue Shenwei saw a photograph and a rubbing of this second bone with their owner but it was too expensive for him to acquire.
c. 1940	Ke Yanling died ('after several more years'). Xue Shenwei bought the second bone from Ke's widow through Xiao Shoutian. By this time therefore Xue Shenwei owned both bones.
1966	During the Cultural Revolution Xue Shenwei is said to have buried the bones in his courtyard for safety. Later he recovered the bones undamaged.
1983	Xue Shenwei indicated that he wished to present the bones to the State Administration of Cultural Heritage, and Jiuan Liu and Nanfang Wang undertook their first study. Lanpo Gu, the director of the Institute of Vertebrate Palaeontology, identified the bones as fossil horse bones. Professors Zhichun Lin and Jingru Wang identified the script of the bones as cuneiform. These invited Chi Yang and Yuhong Wu, both of whom had studied cuneiform, to work on the inscriptions.

1984   Chi Yang identified one bone, here called Bone 1, as written in Akkadian (i.e. Babylonian) language of about 1000 BC, and identified the important words *Sumer*, *Governor-General*, and *four quarters*.

1985   Professor Zhichun Lin visited Xue Shenwei to find out as much as he could. Xue Shenwei drafted an account of the discovery and history of the bones. Xue Shenwei presented the two bones together with other historic relics to the Palace Museum in Beijing, where they are now preserved. In July of that year he died.

1985   Yuhong Wu spent a study period working on Assyriology in Oxford. There Stephanie Dalley recognised the name of Cyrus (*ku-ra-áš*) in the same bone studied by Yang Zhi (Bone 1), and in due course Professor O.R. Gurney identified the whole of the Bone 1 inscription as covering lines 18–21 of the Cyrus Cylinder. Yuhong Wu wrote an account in Chinese, which was later published under the title 'A Horse-Bone Inscription copied from the Cyrus Cylinder (Line 18–21) in the Palace Museum in Beijing', in *Gugong Bowuyuan Yuankan* [Palace Museum Journal], 1987/2: 34–6. This included rubbings and photographs of Bones 1 and 2.

1986   Yuhong Wu delivered a paper on the findings about this bone at the *Compte Rendu de la XXXIII Rencontre Assyriologique Internationale*, which was held in Paris 7–10 July. He also published an English version of his Chinese article under the title 'A Horse-Bone Inscription copied from the Cyrus Cylinder (Line 18–21) in the Palace Museum in Beijing', in *Journal of Ancient Civilizations* 1 (1986): 15–20. This included rubbings and photographs of both Bones 1 and 2. Bone 2, which similarly contains part of the Cyrus Cylinder text, defied interpretation until the present writer worked on it in early 2010.

1987   Shi Anchang annotated and published Xue Shenwei's account of his rescue of the bones under the title 'An Introduction to the New Fossil Bones inscribed with Cuneiform, Collected by the Palace Museum', in the journal *Gugong Bowuyuan Yuankan* 1987/2: 30–33. A shortened English translation of this article with notes by Yang Zhi was then published under the title 'Brief Note on the Bone Cuneiform Inscriptions', in the *Journal of Ancient Civilizations* 2 (1987): 131–4.

c.1987   At much the same time Xueliang Ma of the Minzu University in China invited Guoyi Luo to examine the bones, and he concluded that they were written in the ancient language Loloish, producing a translation of the text on that basis; see Shi Anchang, *Gugong Bowuyuan Yuankan* 1987/2: 30–33.

Previous interpretations

As clearly explained by Wu Yuhong, Bone 1 is riddled with serious textual errors and omissions. The errors involve wrongly copied signs, words split over the ends of lines, parts of lines in the wrong place and the complete omission of very many signs. These characteristics, together with the lack of any kind of parallel at all for cuneiform inscriptions on bone in China, led him to state:

> I believe, therefore, that this inscription was copied by someone who had access either to a publication of the BM cylinder [or] to the cylinder itself, which is on display in the British Museum. It is also possible, although quite unlikely, that it was copied in late antiquity by someone who found a duplicate of the Cyrus Cylinder and then copied a section of the text, for what reason we cannot be sure, but perhaps even for the magic power that these strange signs may have been thought to possess. (Wu Yuhong 1986: 17–18)

Nevertheless, for the benefit of the workshop discussion the points raised by Wu Yuhong were considered in a working paper by the present writer:

1. *By copying a publication of the cylinder*

    That the inscription could have been produced by an individual who had access to a publication of the cylinder was unconvincing. For all its fame and significance, the actual cuneiform text of the Cyrus Cylinder has remained poorly published, its history being limited – as we then thought – to a copy in cuneiform type in 1880 and an ink drawing made from that in 1890. Leaving aside the question as to whether either of these publications could ever realistically have been available in China, there seemed to be serious difficulties in the idea that the text on Bone 1 was copied from either publication. Although the choice of signs used to spell out the words on the bone does not differ from those used in the Cyrus Cylinder itself, the shape of certain signs demonstrates that some source other than the two publications must have lain behind the bones. In addition, the cuneiform wedges in the bones exhibit a characteristic found neither in the Cyrus Cylinder nor in the published copies of it (nor in replicas of the Cylinder). This feature is that the 'heads' of the component wedges are carved with a notch, like the end of an arrow, whereas ancient cuneiform on clay – as exemplified by the

Cyrus Cylinder itself – is written in wedges with flat or straight tops. At the same time, the stylistic arrangement of the wedges does correspond to real sign forms that can be found in other cuneiform texts.

2. *By copying the Cylinder itself in the British Museum*
   That a forger produced the bones by copying directly from the original cylinder itself can be ruled out, as it is known in the British Museum that this has never happened.
3. *By copying a duplicate of the Cyrus Cylinder in late antiquity*
   Replica Cyrus Cylinders, of which an abundance is in circulation and whose standard of reproduction varies from excellent to illegible, cannot have furnished the inspiration for the bone texts.

Furthermore, Bone 1 exhibits a textual variant, writing the real signs IGI.MEŠ ('foremost') instead of *ù* ('and') (written IGI.LU). Since this is meaningful in the context and involves a real cuneiform sign form as an alternative, it has seemed hard to maintain that the bone is nothing more than a modern forgery. Who could possibly have been responsible?

It has to be recorded that the workshop discussion was unable to circumvent these specific points even though the taint of forgery remained largely undispelled. The simplest explanation in the circumstances seemed to the present writer to be that the two bone texts from China represented defective and part-copies of a lost original monument inscribed with the same text as that found on the Cyrus Cylinder, which were not necessarily designed to mislead. Such an explanation would account for all considerations described above and could stand up even if the bone inscriptions themselves were ever shown to have been produced in relatively recent times, for the problems in identifying what could have been copied would still apply, and a late date of 'production' would not in itself disrupt the possibility that the ultimate source was an ancient and unknown copy. Ideas (now redundant) were therefore mustered to explain how this putative document could have reached the eastern outskirts of Cyrus' empire and how the bizarre patchy nature of the texts could have come about.

Demonstrating forgery

At the very last minute, however, a completely overlooked publication of part of the Cyrus Cylinder inscription was finally tracked down which

> THE CUNEIFORM ACCOUNT OF THE CAPTURE OF BABYLON BY CYRUS,
> WITH HIS GENEALOGY. Lines 15–21. (See page 85.)

overturned instantly all the argumentation that could be mustered in defence of the bones. In 1884, one year after he joined the staff of the British Museum, E.A. Wallis Budge published the first edition of his popular work *Babylonian Life and History* (Ismail 2011: 57). This volume contained an early photograph of the Cyrus Cylinder, a long translation summary of the cuneiform text on pp. 79–82, and an extract in a rather idiosyncratic cuneiform font (*figure 11*).

This font had been available as early as in Rawlinson 1846, continuing in use for nineteenth-century journals, such as Rawlinson 1880, or the London-produced *The Babylonian and Oriental Record*. A superior font, however, was employed in the official publication of the Cyrus Cylinder by Sir Henry Rawlinson and Theophilus Pinches in 1880, as well as, for example, in Budge 1880. Crucially, the Cyrus Cylinder extract given in Budge 1884 is in the older font. Comparison clarifies the following telling points:

1. Budge's (1884) cuneiform font exemplifies the very characteristic found so troublesome above, that the heads of the upright and horizontal wedges display a notch like the end of an arrow.

11 The forger's crib; lines 15–21 of the Cyrus Cylinder text as given in E.A.W. Budge's *Babylonian Life and History* (1884: 80).

32   THE CYRUS CYLINDER

2. Budge's published extract from the Cyrus Cylinder text runs from lines 15 to 21.
3. Bone 1 has 15 lines of 'text' with assorted signs from Cyrus Cylinder lines 15, 17–21 and 27, 28+.
4. Bone 2 has 15 lines of 'text' with assorted signs. Of these lines 1–7 cannot be identified as being from the Cyrus Cylinder; lines 8–13 cover Cyrus Cylinder lines 8–9, 11–12, 15–21, 23–6; lines 14–15 cover Cyrus Cylinder lines 35+.

The overlap between the Budge extract and Bone 1 is compelling. Both come in at line 15, but Bone 1 also contains a small number of signs from beyond line 21 not in Budge. Bone 2 begins with unidentifiable material and covers lines 8–12 with no support from Budge; like Bone 1 it covers Budge's lines 15–21 but continues from 23 to 26 and even beyond line 35 with no support from Budge.

Comparison of the signs in the text of Budge (1884) with those in the two bones leaves no doubt that the latter derive from and are dependent on the former. It looks as if someone took the printed lines and made them especially difficult to understand according to several clear processes: omission of many signs, splitting words and so on, as indicated above. If so, it is important to explain how the bones can include material that is not in the Budge plate. The only plausible explanation is that Budge, then a keen student of cuneiform, made up his own complete copy of the 45-line Cyrus Cylinder inscription using the old cuneiform font soon after his arrival in the Museum in 1883. In 1884, however, Budge was, rather resentfully, warned off cuneiform research – the reserve of the superior scholar Pinches – by the Museum administration, and was instructed to stick from then on to Egyptian hieroglyphs (Ismail 2011: 58). One can only assume that whoever produced the two bones somehow had access to Budge's complete copy of the Cylinder, while a 15-line extract sufficed for his book. The Cylinder photograph and specimen text in cuneiform font appeared in at least three of the four printings of the first edition of Budge's *Babylonian Life and History*, but both were omitted in the revised and rewritten second edition, which eventually appeared in 1925.

It is pointless to speculate who might have been responsible for inscribing the fossil bones, or why the text is so garbled and incomplete. We know nothing of when or where they were produced, and what might have happened thereafter. The identification of the texts is far from straightforward, especially with Bone 2, which was not in fact 'deciphered' until 2010! Whether idleness, financial greed, malice or revenge lies behind them cannot be established. Perhaps one day some other piece of the story will emerge.

The important point is that, despite many unknowns, it is now established that the two fossilised horse bones that came out of China to perplex the world of Assyriology are undoubtedly worthless forgeries and can be entirely dismissed from Cyrus Cylinder studies from this point on.

2

The Cyrus Cylinder: discovery

JONATHAN TAYLOR

The discovery of the Cyrus Cylinder

On 17 November 1879 at a meeting of the Royal Asiatic Society in London, Sir Henry Creswicke Rawlinson read a paper on 'A Newly Discovered Cylinder of Cyrus the Great', which he described as 'the most interesting historical record in the cuneiform character as yet brought to light'.[1] *The Times* report continues:

> It was not among the monuments lately brought home by Mr. Hormuzd Rassam himself, but must be credited to his last archaeological explorations in the East, under the auspices of the British Museum, having been sent to this country by one of the agents left behind by him to continue his excavations in the Mesopotamian mounds. It is in the Babylonian script, as was to have been expected from its having been discovered among the ruins of the Birs Nimrud, the acknowledged site of the ancient Borsippa, of which city, as Sir Henry Rawlinson remarked, it was the more surprising that it makes no mention.

So began a long history of fascination with, and confusion over, this iconic object.

Rawlinson's comments drew a stern response from Rassam. He wrote indignantly to Samuel Birch, keeper of the then Department of Oriental Antiquities at the British Museum, on 20 November:

> The Cylinder of Cyrus was found at Omran[2] with about six hundred pieces of inscribed terracottas before I left Baghdad. There was another Cylinder

which was found at Birs Nimrud which is I believe with the collection in Mr Pinches' room. I am sorry that Sir Henry Rawlinson did not ask me about it before he read his paper before the Asiatic Society because he would not have then made that mistake about the time and place of its discovery. I <u>stopped</u> the work at Birs Nimrud <u>before I left Baghdad</u> and as I <u>never</u> brought any collection <u>with me</u>, but all my discoveries followed me as in the case of the Cylinder and the 800 pieces of tablets. We have not yet received any collection of antiquities which were discovered <u>after my leaving Baghdad</u>. If Colonel Rawlinson can wait until I come to the Museum on Monday I shall be able to find out more correctly which of the cylinders was found at the Birs. (ME Corr. 5301)

12   Sir Henry Creswicke Rawlinson, who broke the sensational news of the Cylinder's discovery.

Anonymous engraving, c. 1873.

In the print publication of his paper Rawlinson neglected to acknowledge the correction, writing instead:

I have not been able to ascertain the exact spot where this Cylinder was found. It is understood at the British Museum to come from the excavations at Birs Nimrud, but I can hardly believe this to be possible, as there is no allusion to Borsippa or to its temple in the whole extent of the Inscription; I should rather judge from the context that it must have been deposited in the lesser shrine of Merodach on the 'holy mound,' which, as I have before said, is represented by the ruins about the tomb of Amrán. (1880: 83)[3]

If we were to accept Rassam's account, then it would be possible to deduce through his correspondence with the Museum that the Cylinder must have been found some time between 10 February (when excavations at Babylon began) and 24 March 1879 (when Rassam left Baghdad). There are, however, problems with his version of events too. Even in the year of its discovery, the Cylinder was already the subject of much confusion.

Rassam, the Cyrus Cylinder and what really happened

There are two principal questions about the discovery of the Cyrus Cylinder: when and where it was found. In both cases there is evidence on paper, but this is tantalisingly incomplete and it is still not possible for us to reach complete certainty on either point. The following paragraphs detail the circumstances of the discovery, as far as they can be ascertained from published sources and the archives of the British Museum. The following paper resources shed light on this complex of enquiries:

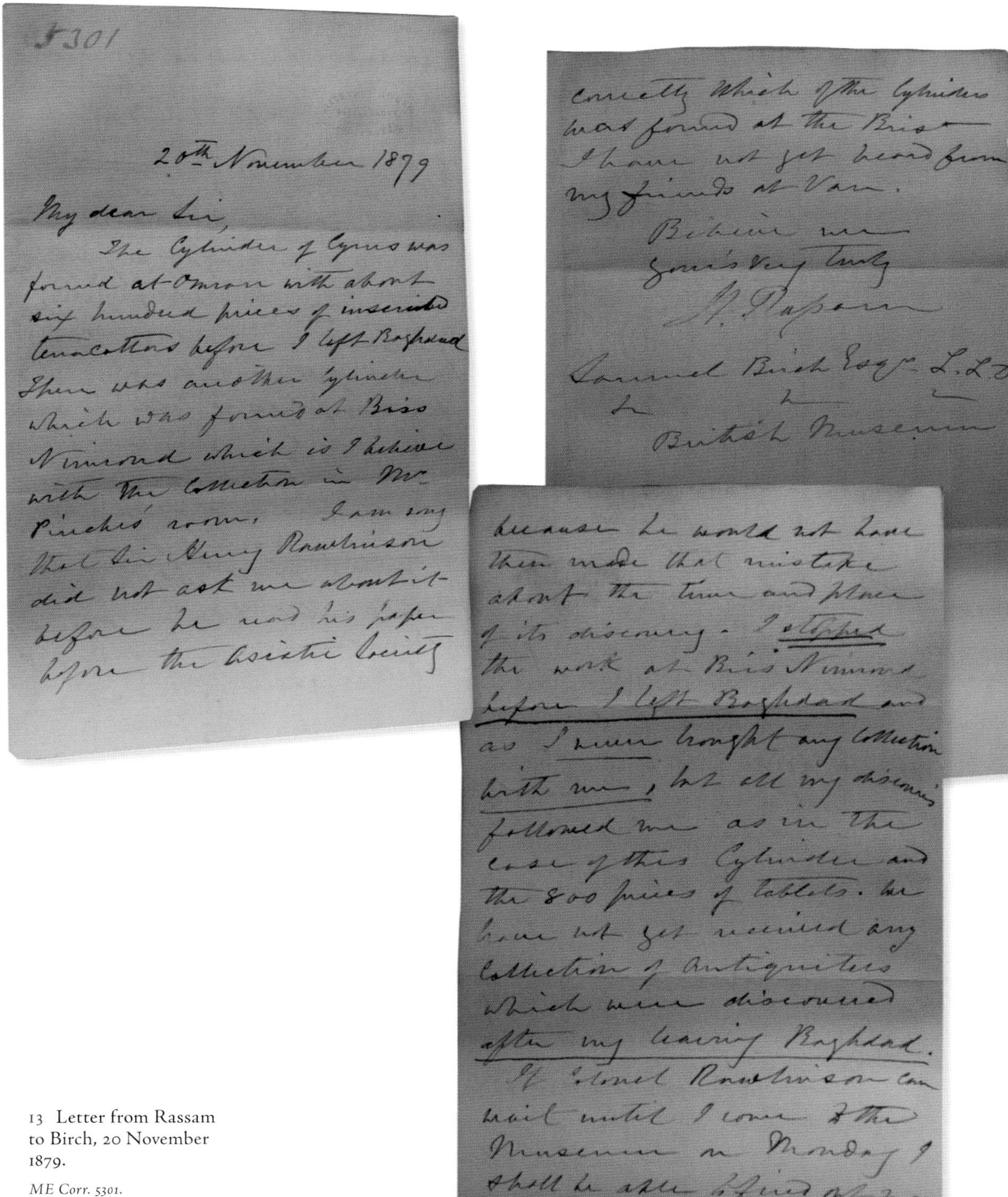

5301

20th November 1879

My dear Sir,

The Cylinder of Cyrus was found at Omran with about six hundred pieces of inscribed terracottas before I left Baghdad. There was another Cylinder which was found at Biss Nimroud which is I believe with the Collection in Mr Pinches' room. I am sorry that Sir Henry Rawlinson did not ask me about it before he read his paper before the Asiatic Society

because he would not have then made that mistake about the time and place of its discovery. I stopped the work at Biss Nimrod before I left Baghdad and as I never brought any Collection with me, but all my discoveries followed me as in the case of this Cylinder and the 800 pieces of tablets. We have not yet received any Collection of Antiquities which were discovered after my leaving Baghdad. If Colonel Rawlinson can wait until I come to the Museum on Monday I shall be able to find out more

correctly which of the Cylinders was found at the Biss. I have not yet heard from my friends at Van.

Believe me
Yours very truly
H. Rassam

Samuel Birch Esq. L.L.D.
&c &c
British Museum

13 Letter from Rassam to Birch, 20 November 1879.

ME Corr. 5301.

- Autograph letters from H. Rassam to S. Birch at the Department of Oriental Antiquities and to successive principal librarians at the British Museum, John Winter Jones (until August 1878) and Edward Bond (from August 1878).
- Letters from H. Rawlinson to S. Birch.
- Letters from S.B. Miles, consul in Baghdad, to the principal librarian.
- Telegrams from Rassam to the principal librarian.
- Despatch inventory by Daud Toma, Rassam's assistant and overseer of excavations.
- Receipt inventories from London, by curator T.G. Pinches, submitted by keeper S. Birch to the Trustees of the British Museum.
- Abstracts of Minutes of the British Museum Trustees' meetings.
- Annual Returns of the Trustees to the House of Commons.

Correspondence is to be found in the Departmental Correspondence volumes (ME Corr.) and in the Original Papers (OP; for correspondence with the principal librarian) stored in the Central Library; further letters from a private collection (PC) have also been made available to me (some of this material has been included in Razmjou 2010).

Background to the excavations at Babylon in 1879

The Babylon of the Bible was an enduring source of fascination. By the mid-nineteenth century a long series of European travellers had begun exploring the site, searching for the remains of the fabled Tower of Babel and the Hanging Gardens of Babylon. In the 1810s Claudius James Rich, the East India Company's Resident at Baghdad, initiated a more scientific study of the site. His *Memoirs on the Ruins of Babylon* (1815; 1839) give us the first accurate topographic map of the site (drawn by Captain Lockett), and the tablets, cylinder seals and other objects he found there would form the nucleus of the British Museum's Mesopotamian collection. Further British excavations took place under Captain Robert Mignan of the East India Company during the 1830s, under Austen Henry Layard in 1850, and – following a brief campaign by Fresnel and Oppert for the French in 1852 – under Rawlinson in 1854. In 1859 William Beaumont Selby, a naval officer operating along the Tigris, surveyed Babylon and produced a new

14 Hormuzd Rassam, who led the British Museum's excavations at Babylon in 1879. Oil painting by Ackland Hunt, 1869.

BM 1955,0630.1.

map, updating that of Rich. Despite all these efforts, Babylon yielded little to rival the spectacular remains from Assyria.

Hormuzd Rassam had been Layard's assistant during the British Museum's excavations in Iraq 1845–51. Following Layard's move into politics, Rassam was given control of the excavations. In the 1870s he was busy working in Assyria. But he was keen also to resume British research at Babylon, spurred on by news that a great many cuneiform tablets were then being found there. Large numbers of tablets and other objects from Babylon had come up for sale on the antiquities market in Baghdad. In 1876 British Museum curator George Smith bought a group of tablets that had been dug up by locals at Amran/Jumjuma the previous winter. Rassam himself bought a collection of seven Babylonian inscribed clay cylinders in December 1877 (OP 913). Convinced by the importance of the objects and mindful that as a Museum employee he could not collect them in competition with the Museum, he stresses that should the Museum not want the duplicates, he would buy them himself (OP 3324; 29 April 1878, letter to Birch). He soon realised that the dealer Marini had many more on offer: 'There is another collection of about 100 inscribed cylinders found at Babylon' (OP 913; 24 December 1877, letter to Birch). In May 1879 the British Museum Trustees purchased from another dealer, Shemtob, 1,500 tablets from Babylon (Minutes 1206).

Before Rassam could start digging at Babylon, however, he needed permission from the Ottoman authorities, who then administered Mesopotamia, and from his employers at the Museum. His old mentor, Layard, now ambassador in Constantinople, described the former as presenting him more difficulty than did negotiating possession of Cyprus (OP 6326). The response from the Museum clearly took longer to obtain than Rassam had hoped. Already on 24 December 1877 Rassam had written to John Winter Jones, principal librarian of the British Museum:

> I had hoped that I should be able to have some workmen at Babylon before I went to Mossul for the purpose of searching for inscribed tablets where I believe a large quantity has been found. But it appears to me that I am limited in the Firman[4] to a certain locality and will not be allowed to make any explorations anywhere else. (OP 687)

The next month Rassam wrote again, ruefully:

> I find that the Firman gives us permission to excavate at Babylon and had this been sent in the telegram which reached the Consul General from Constantinople before I left Baghdad I would have left a few workmen to excavate at the spot where I was told heaps of terra-cotta inscribed tablets have been found. (OP 1486; 12 January 1878)

The location of this bountiful spot is spelled out in another letter later that month:

> I was very anxious while I was there to place a few workmen at a place called 'Alkasir'[5] where I was told a large number of tablets had been found, and of which I had purchased a few fragments. (OP 1681; 26 January 1878)

In March 1878 he wrote to the Museum:

> I also telegraphed[6] for permission to send a trustworthy agent to Babylon to dig for inscribed terracottas as I was told that they were to be found in a great quantity near Hilla.[7] It is desirous to keep the Arabs from destroying these valuable historical relics until my return if I should come out again next autumn.[8] (OP 2636; 23 March 1878)

His efforts were in vain; on 22 April he wrote with growing frustration:

> With regard to Babylon as I received no permission from the Trustees to send an overseer with a few workmen to dig there according to the intimation in my telegram to the Principal Librarian of the 20th March of course I did not do so. (OP 3051)

On 24 July he persisted:

> I would recommend that in my next expedition to the Valleys of the Euphrates and the Tigris I should be allowed to devote the grant allowed me half for the excavations in Assyria and the other half for the explorations in different parts of Mesopotamia especially in and around Babylon where I hope to find important collection of inscriptions. (OP 4314)

Before setting out for the autumn season, he repeats:

> I propose that ... I should proceed to Babylon about the middle of December and explore some parts there for inscribed terracottas. (OP 5396; 2 October 1878, letter to Winter Jones).

Rassam's task was complicated by a bout of ill health, interruptions to the postal service, and the difficulties of negotiating permission and financial details with London. But in January 1879 Rassam received the new principal librarian's 'telegram of the 23rd in which you authorise me

DISCOVERY 41

to proceed to Babylon, as I proposed, and I shall therefore go down to Baghdad by raft next week and hope to be able to commence operations near Hillah about the 10th or 12th Proximo [i.e. February]' (OP 1218; 25 January 1879). After more than a year of delays, Rassam could at last explore Babylon.

The spring 1879 season at Babylon

Rassam left Mosul on 30 January 1879 (OP 1465) for the journey by raft to Baghdad, stopping en route at other excavations to make arrangements for the work there to continue. With him travelled Daoud Toma, his trusted overseer who had worked with him in Assyria, and Ahmed al-Abid, an experienced digger whose services he had obtained by chance as he was leaving Baghdad (Rassam 1897: 259). The party reached Hillah, the town adjacent to the site of Babylon, on Monday 9 February (OP 1184). His first order of business was to visit the lieutenant governor, Mohammed Pasha, an elderly Kurd 'of the old school, who considered searching for antiquities a silly occupation, and those who valued them were only fit for a lunatic asylum' (Rassam 1897: 259). He was unable to get an appointment on the first day because Mohammed Pasha feigned illness. The next day the weather was wet, so the lieutenant governor had to stay in his harem; the rumour was that he had just received a pretty girl from Constantinople. Rassam sent the *firman* and local passport ahead, with a message that he could not afford to delay; a reply came to say that he was to dig as he wished. The two men only met a couple of days later. Rassam continues the story:

> as soon as I arranged with the authorities about commencing researches in Babylon I engaged the required number of workmen and placed different gangs to excavate in six localities; Viz. the mound called Babel on the supposed site of the hanging gardens, Alhimaira at the south-eastern entrance of the city, Imjaileeba on what is commonly called the Kasir or Palace, Omran, Jimjima, and Birs Nimrud. Babel is at the extreme northern extremity of the supposed city of Babylon and Jimjima is at its southern end, and Birs Nimrud stands about ten miles to the south. (OP 1465)

A crucial problem that plagued archaeology at this time was illegal digging. It had long been the practice for locals to dig bricks out of sites for use in modern construction projects.[9] Inevitably in the course of such digging antiquities were uncovered. Rassam explained to Winter Jones:

> I think it is most important that we should devote a small portion of the grant on researches in Babylon where it is likely we may find some valuable records. The Arabs who are bribed [locally] to dig clandestinely for Babylonian remains break everything they find, partly from ignorance, and partly for the sake of increasing the number of the pieces which they sell to different people. Generally speaking, an Arab digger contracts with two or three individuals to provide them with a certain quantity of antiquities, and when he cannot supply each individual with the same number or quality of objects, he breaks a most valuable inscription to divide amongst them.
> (OP 5396; 2 October 1878)

These would be sold for a few pennies to dealers, who would then sell them to western collectors at vastly inflated prices; Rassam (1897: 264) tells of seeing tablets from Babylon for sale in Baghdad at prices over a hundred times what had been paid to the diggers. He lamented that

> The damage done by such mode of searching is incalculable, inasmuch as the Arab style of digging is too clumsy to get out fragile objects intact from narrow and deep trenches, especially when they have to carry on their work as secretly as possible, from fear of being detected by the authorities. In nine cases out of ten, they break or lose a large part of their collections, and worse than all, they try to make a good bargain by breaking the inscribed objects, and dividing them amongst their customers. (1897: 262)[10]

This phenomenon, widespread across the world at this time, has implications for the modern analysis of cuneiform archives. The baksheesh system was later invented to counter it. Rassam explained to Birch his tactics for combating looting as follows:

> The Ottoman government has issued stringent orders against those who excavated for antiquities without special notes from the Porte but there are a dozen of ways of evading those orders especially through the connivance of the local authorities. My first case on reaching Hillah was to engage those Arabs in my service who used to dig for bricks in Babylon; and it may appear to you strange that I depend mostly upon these very men to keep [the local dealers] from committing unlawful acts.[11] I pleased these men greatly by allowing them to have besides their pay every brick they found in their diggings excepting bricks with inscriptions. Had I not done this they might have refused to work with me and yet went on with these excavations under the plea that their forefathers were allowed to do so from time immemorial as all the towns and villages in the neighbourhood have been built with materials obtained from the ruins of that ancient city.
> (ME Corr. 5297; 24 February)

Looting continued, however, fuelled by the dealers in Baghdad. Rassam (1897: 263–4) explains that he decided not to bring any prosecutions because he lacked definitive proof, and was inclined to overlook minor misdemeanours rather than create difficulties with the local authorities. These challenges aside, results came almost instantly from the excavations at Babylon:

> I am glad to report that I had not long to wait before I was rewarded by the discovery of inscribed terracottas both in the city of Babylon and Birs Nimrud, and I have great hopes that once I return to Mossul I shall be enabled to send to England some valuable collection of inscriptions. But I beg to bring to the notice of the Trustees of the British Museum that the different mounds on the site of Babylon have been for centuries so much searched into for bricks and other materials by the natives of the country without order or method that nothing but heaps of rubbish can now be seen scattered over the place, and it requires immense labour and great expense to get at the remaining spots which have not been trashed by unskilful hands. My first aim was on commencing the excavations to get at the walls of some chambers to enable me to direct the labourers to carry on the work properly and regularly, because I feel confident that the Arab workmen never take the trouble to dig to the bottom of the chambers when some valuable relics might yet be found. We have now come upon the walls of chambers built of sun-dried bricks both in Jimjima and Birs Nimrud, and by following these walls I hope to find my unaffected ground. (OP 1465; 16 February)

Rassam himself would not stay long at Babylon. Reports had reached him that Telloh was a site worth investigating. On 16 February he wrote to the Museum:

> I intend, therefore, to leave this [Hillah] for Baghdad on the 19th instant [February] so as to proceed thence to the Hai river by a steamer, as I find it would be the shortest way to reach the mound I wish to examine. I shall leave the same number of workmen to excavate in Babylon and Birs during my absence under the superintendence of the overseer I have brought with me from Mossul [Daoud Toma], and if I find before I leave this country that we cannot afford to keep so many hands at work during the month of March I shall only keep three gangs here to retain the right of research in the most attractive localities. (OP 1465)

On 24 February Rassam left Baghdad for Telloh (OP 1904), hoping to return to Baghdad in ten or twelve days. As for the situation at Babylon meanwhile: 'I sent you a telegram on the 21st and informed you of my return to Baghdad from Hillah and of my having left about 80 men excavating in different parts of Babylon. There is a great field for research there and I do not think we ought to abandon it for the benefit of others.' He added to Birch:

> When I left Babylon on 19th these were finding a good number of inscribed terracottas in two of the trenches, but unfortunately they are fragments. Yesterday I received an important telegram from Daoud our overseer in which he mentioned to me signs that he had found a 'marble chair' at the Birs Nimrud that at 'Jimjima' and 'Omran' they were finding what I am 'reckoning upon'. Had I not [an] important mission on the river 'Hai' [i.e. the trip to Telloh] I would have pressed at once to Babylon. But as I expect the details in a letter on my return to Baghdad I would certainly visit the place again before my return to Mossul. … Hope that I shall be able to dispatch you some important and valuable antiquities on my return from the ruins of Babylon. (ME Corr. 5297; 24 February)

On Wednesday 19 March Rassam wrote to the Museum from Baghdad, reporting his work at the site of Telloh up to his departure from there on the 6th (OP 1905). He also provided an update on work at Babylon: 'Both in the City of Babylon and Birs Nimroud our workmen are finding inscriptions.' He noted his intention to leave for Mosul before the end of that week, but 'I am only waiting to receive and dispatch to England all that has been discovered in Babylon' so that these could be sent together with those found at Telloh. Rassam handed the consignment over on Monday 24 March (OP 2101). According to his inventory it included tablets, bricks, vessels and other objects from Babylon and Birs Nimrud as well as Ashur and Telloh.[12]

The role of superintendent of excavations would pass to Colonel Nixon, Consul in Baghdad. On 22 March Rassam wrote to him to explain what this role would entail. The overseer at Babylon, Daoud Toma, was to render account to Nixon at the end of each month. Rassam meanwhile reduced his workforce:

> There will be three gangs of workmen of 7 men each digging in different parts of Babylon, 18 of whom receive three Piastres each per day, and 3 at four Piastres each – over these workmen there are two overseers, one is Dawood Toma, who receives 300 Piastres a month with 100 Piastres for allowances, and the other overseer, Sayad Ahmed Alahid, receives 150 Piastres. (OP 2101)[13]

Rassam then reported his actions to Edward Bond, who had replaced Winter Jones as principal librarian of the British Museum, on 24 March

(OP 2101): 'I have made over charge of the excavations in Babylon to Colonel J.P. Nixon, Her Majesty's Consul General at Baghdad, and herewith I beg to append a copy of my instructions to him.' Work would go ahead as was planned.

> I have been obliged to prolong my stay here on account of the dilatoriness of the Turkish authorities in examining the antiquities I wish to send home. I hope, however, to be able to leave Baghdad today for Kala Shirgat and Mossul … I have received a telegram from our overseer at Babylon of his having discovered about 700 pieces more of inscribed terracottas which he is sending to Baghdad; but as this collection will not arrive here till after my departure I have asked Colonel Nixon to forward them to you as soon as possible after he receives them.

Meanwhile Rassam's instructions to Nixon to settle the accounts monthly had caused problems back in Babylon. Rassam wrote from Mosul on 28 April:

> I am also in trouble about my excavations at Babylon as our overseer there informs me that Colonel Nixon refuses to pay the workmen weakly [*sic*] which has always been the case and says he would only pay them when the month is over. This obliges the overseer to retain bad workmen as of course if he wants to dismiss them he has no money to pay them. (ME Corr. 5298, letter to Birch)

Nixon's position is understandable, since Rassam had instructed him that accounts would be rendered monthly. The problem must have been solved, however, since on 6 June Rassam wrote to the principal librarian (OP 2907) from Smyrna as follows: 'I am glad to say that from a report I have received from Baghdad since I left Mossul I learnt our excavations in Babylon are still progressing very satisfactorily, and that a large number of inscribed terracottas have been found there lately.' On 14 June Birch (OP 2909) reports to the principal librarian that the results from Assyria were so disappointing that he was thinking of focusing on Babylonia. By August a new consul general, Samuel Barrett Miles, had taken responsibility for the excavations at Babylon.

Once Rassam arrived back in England, he announced his decision to resign his post, on account of family affairs (OP 2948, letter to the principal librarian). Work was continuing in Iraq, but now the Museum had to find a replacement for Rassam. Birch reported his resignation to the Trustees on

27 June (OP 3091): 'The explorations in Babylonia and the neighbourhood of Bagdad will require European supervisal and Dr Birch has no one whom he could suggest at present for that last which is too remote and extensive to be compatible with the duties of the consular authorities at Bagdad.' He sought an honorarium for Rassam, to offset his expenses. Rassam was to prove a hard man to replace, however. Layard wrote to Birch (ME Corr. 3231; 14 July) declaring that he had no idea who could replace him. The same day he wrote to the principal librarian: 'From his tact, experience and acquaintance with the languages and customs of the people, he is singularly well qualified for the work that had been assigned to him. I am quite at a loss to recommend any one to succeed him' (OP 3284). Rassam would prove so difficult to replace that the Museum eventually had to persuade him to resume his work in Iraq. He went on to lead a series of expeditions over the next few years, and came to be responsible for finding many of the most important Mesopotamian objects in the British Museum collections (Reade 1993).

When was the Cyrus Cylinder found?

Rassam never claimed to have found the Cylinder himself, and at no point does he mention to any of his correspondents in the British Museum that an object identifiable now as the Cyrus Cylinder has been found. This is despite several reports of similar objects (prisms and barrel cylinders of Assyrian kings) found throughout the archives and a preoccupation with cylinders in general as particularly valuable sources for reconstructing Mesopotamian history. Recent opinion asserts that the Cylinder was discovered between February and March 1879.[14] This is on the basis of Rassam's letter of 20 November (for which see pp. 35–6, above), which claimed that no antiquities found after his departure from Baghdad had yet arrived. Rassam's departure from Baghdad is certainly dated to the end of March 1879, as shown in the timeline of his movements (*table 2.1*):[15]

The despatch inventory included in OP 3984 records that the antiquities which arrived at the British Museum in August 1879 included some excavated as late as 12 April. Thus Rassam was mistaken on this point of chronology; objects excavated after his departure from Baghdad had indeed arrived in London already. We are left to establish the circumstances of

TABLE 2.1 Timeline of Rassam's movements

10–19 February	Rassam at Babylon; during this time he moved between the Babylon sites and Borsippa (for which see e.g. Rassam 1897: 268). On 19 February he left for Baghdad.
21 February	Rassam arrived in Baghdad.
24 February	Rassam set out for Telloh, hoping to return to Baghdad in ten or twelve days. He wrote that he planned to go back to Babylon between his return from Telloh and his departure for Mosul. The trip to Telloh took much longer than he had expected, however, due to unseasonal weather. He left Telloh on 6 March.
14 March	Rassam arrived back in Baghdad (OP 1905; Rassam 1897: 284). The time between then and his departure was filled with making arrangements for the discoveries from Babylon to be examined by Turkish officials before shipping to England, and setting Colonel Nixon in charge of continuing the excavations (for details of which see Rassam 1897: 284–5).
19 March	Rassam wrote from Baghdad to say that he was waiting to receive and dispatch to England all that had been discovered in Babylon (OP 1905).
22 March	Rassam handed over responsibility to Nixon (OP 2101). There is no evidence to suggest that Rassam ever actually managed to return to Babylon during this trip, and indeed there seems not to have been sufficient time for him to have done so.
24 March	Rassam passed the recently discovered antiquities to Malcolm Baltazar, an Armenian merchant (OP 2101). He then left Baghdad, northbound.
2 April	Rassam reached Mosul (presumably via Ashur), and left there on 2 May (OP 2534). By 6 June he was at Smyrna (OP 2907).

the find as best we can through the archives. The lack of detail in these records prevents complete correlation with the extant tablet collection at this stage.

We can correlate the 1879 shipments of antiquities using dispatch reports to the Museum and the documentation concerning their arrival and processing at the Museum; evidently the volume of material was so large that the whole process took several years to complete. Three shipments are relevant to the present discussion, as outlined below. A complication is that tablets purchased from dealers also arrived at the Museum during this time, including a batch originating with Marini, bought by the Museum from Spartali in March (the 'Sp 1' collection), a second from Shemtob was authorised in May (becoming part of the 1881,0706 collection), and another purchase from Spartali, which was in the Museum by August (the 'Sp 2' collection). For an account of the shipments and the processes of packing, checking and recording, see Reade 1986; note that tablets excavated by Rassam in 1878–79 were registered as the collections Rm I–IV.

Details of shipments

SHIPMENT 1

Rassam handed over to Baltazar three cases and two packages on 24 March (OP 2101); these were actually dispatched on 15 April (OP 2325). The contents of this consignment were tablets, figurines, vessels, bricks and other objects, mainly from Babylon and Birs Nimrud, but also Assur and Telloh. They reached the British Museum on 21 June, where together with a stray box from the 1881,0324 collection they would later be registered as the Rm III collection (= 1879,0620). They were presented to the Trustees on 12 July (OP 3282). The inventory drawn up in London includes entries for:

> 110. Part of Inscribed Cylinder from Birs-Nimroud.
> 111–159. Fragments of Tablets from Birs-Nimroud.
> 160–297. Fragments of Tablets. 'Jimjima,' Babylon.
> 298–634. Fragments of Tablets. 'Omran,' D[itt]o.
> Several very small fragments of tablets for joining.

Item 110 is BM 33428 (Frame 1995: 123–6; B.6.14.2001), the cylinder from Borsippa that caused confusion for Rawlinson, much to Rassam's annoyance. A fragment from 1880,1112 was later joined to it.

SHIPMENT 2

After Rassam's departure, Toma continued excavating. Among the finds discovered very shortly thereafter was the Cyrus Cylinder. Toma drew up an inventory, sealed it, and sent it with the objects to the consulate in Baghdad. There a clerk prepared a slightly abbreviated English translation of it for London (OP 3984). It states:

List of antiquities which were discovered at Telambran and Jumjamah from 17th to 23rd March 1879.

Tablets with writing on, large and some of them unbroken		50
d[itt]o	four fingers wide	110
do	three fingers wide	260
do	small	390
Cylinder unbroken[16]		1
Fragment of a pin and a dog's head of bone		2
Bricks with figures and designs		6
		819

The inventory confirms Rawlinson's public account that the Cylinder was among objects recently discovered by an agent left by Rassam to continue excavations. Rassam's absence from Babylon at the time of discovery is unfortunate, since he would have recognised the Cylinder later, and given us a clearer idea of its findspot. His correspondence shows that he took an active interest in inscriptions, and was even able to produce reasonable drawings of them. At the consulate Toma's cases were unpacked, cleared by the Turkish representative, and repacked for shipping.[17] Rassam reported to Birch with alarm on 28 April that:

> I heard also that the large quantity of the inscribed tablets which were found in Babylon where [sic] kept without them being sent on to England as I had requested him [consul Nixon] to do – so I had to telegraph to him about them as I was afraid that he would keep them with him for ever! (ME Corr. 5298)

A local merchant, Baltazar, was finally entrusted to ship them by steamer via Basra. Shortly after their arrival at the British Museum in August, they were inventoried by Pinches, independently of the dispatch inventories (which were sent separately and to the central administration). Pinches' brief inventory was presented to the Trustees on 3 October (OP 4228); it contained the following section:

> Inventory of tablets, &c, excavated by Mr Rassam's overseer at Babylon & Birs Nimrud, received in August last…
> Clay Tablets from Babylon
> 33. Portion of a fine cylinder containing an account of Cyrus' taking of Babylon, his genealogy, &c, &c.
> 34. A broken Tablet containing a list of stones.
> 35. Part of a four-column syllabary.
> 36. A contract Tablet.

c. 11 Oct. 1879.

List of Antiquities which were discovered at Jelambran and Jumjamah from 17th to 23rd March 1879.—

	Nos
Tablets with writing on, large and some of them unbroken	50.—
do. four fingers wide	110.—
do. three fingers wide	260.—
do. small	390.—
Cylinder unbroken	1.—
Fragment of a Pin and a Dog's head of bone	2.—
Bricks with figures and designs	6.—
	819.—

List of Antiquities discovered from 17 March to 12th April 1879.—

	Nos
Burnt Tablets (unbroken) with writing on	6.—
Unburnt — do — four fingers wide	8.—
Large Tablets unbroken	2.—
Tablet with writings, three fingers wide	47.—
Tablets small and some of them burnt	293.—
	356.—

Miscellaneous Articles

	Nos
A female figure riding on a Bullock	1.—
Figure of two Soldiers	1.—
	figure

15 The dispatch inventory drawn up by Toma in Babylon, 1879, and its contemporary translation.

OP 3984.

37–50. 14 large pieces of Tablets referring to calculations, omens, &c.
51–1032. Fragments of Tablets referring to trade, to mythology, and to mathematics, also fragments of Bilingual lists, &c, &c.

The shipment containing the material listed by Toma is securely correlated with this one presented to the Trustees in October. And a consignment dispatched in May (following Rassam's chivvying letter in April) could indeed be expected to arrive in London early in August; shipments 1 and 3 each took nine weeks to arrive.

Shipment 2 contained six cases (OP 4228), which Miles (OP 3984) described to the Trustees on 20 August as comprising four cases of antiquities sent already by Colonel Nixon plus two more sent by Miles himself. The first four cases would have held the 700 or so tablets about which Toma had telegraphed Rassam while he was still in Baghdad.[18] Rassam recorded in his letter of 24 March that these would not arrive until after his departure (OP 2101). The source of his error in stating in November that nothing excavated since his departure from Baghdad had yet arrived in London is perhaps to be attributed to his not having realised that the shipment that arrived in August contained not only the four boxes he had tasked Nixon with sending, but also two further boxes with material excavated later. The six cases in total were inventoried by Toma as containing about 1,100 fragments. Unfortunately, the registration of this shipment is less clear than that for the previous and subsequent ones. Some tablets and other objects, including the Cylinder itself, were later given numbers in the 1880,0617 collection.

SHIPMENT 3

On 20 August Consul Miles reported that in addition to the six cases already dispatched, a further two would be sent soon. These arrived on 24 December, and would be registered as the Rm 4 collection.

Summary of key points

The Cyrus Cylinder arrived at the British Museum in shipment 2, and the following timeline summarises the preceding events:

1. Toma writes an inventory and seals it as 'Overseer of excavations'.

27. Bronze instrument.
28. Parts of bronze and other ornaments.
29–31. Portions of bone instruments.
32. Fragment of a Babylonian boundary stone containing portion of snake, &c.

Clay Tablets from Babylon.

33. Portion of a fine cylinder containing an account of Cyrus's taking of Babylon, his genealogy, &c, &c.
34. A broken Tablet containing a list of stones.
35. Part of a four-column syllabary.
36. A contract Tablet.
37–50. 14 large pieces of Tablets referring to calculations, omens, &c.
51–1032. Fragments of Tablets referring to trade, to mythology, and to mathematics, also fragments of Bilingual lists, &c, &c.

Clay Tablets from Birs Nimroud.

1033. Case-tablet dated in the reign of Seleucus.
1034. Contract-tablet dated in the 1st year of Artaxerxes.
1035–1039. Five complete Tablets.
1040, 1041. Lists of Wood, broken.
1042–1141. Fragments of Tablets referring to trade, mythology, mathematics, &c., &c.
1142–1145. Fragments of Glazed and ornamented bricks.

With very few exceptions indeed, all the above-mentioned Tablets are of unburnt clay. Many, therefore, have been broken, or otherwise damaged, in transit, by bad packing. Most of the fragments, found as such, are so very small as to be, in their present incomplete state, quite valueless.

16 The receipt inventory compiled by Pinches of the shipment that arrived at the British Museum in August.

OP 4228.

2. Toma's cases transported on mules from Babylon to Baghdad, under the care of Yusuf Antoon Shamas.
3. Toma's cases unpacked at the Consulate in Baghdad, cleared by the Turkish representative, and repacked. Toma's inventory is translated into English, imperfectly.
4. Miles writes to the principal librarian on 20 August and encloses Toma's inventory. He reports four cases sent by Nixon, two by himself. A further two would follow. Toma's inventory covers the first six cases, plus (apparently in another hand) the remaining two cases.
5. Cases transported to London by Baltazar.
6. Cases received at the British Museum as follows: August: six cases; December: two cases. The two cases, apparently unusually, were inspected by British customs at Custom House instead of the Museum. By the time of their arrival, one of the two cases was damaged, and it was surmised that around twenty objects had been lost.
7. Pinches compiles two receipt inventories, one for the six cases, another for the two. These were submitted to the Trustees in October 1879 and January 1880, respectively.
8. The only cylinder recorded in either of Pinches' receipt inventories is the Cyrus Cylinder, within the August/October shipment.

How an unbroken cylinder became broken

The cylinder described as 'true cylinder without breaks; small' in Toma's dispatch inventory must correspond to the Cyrus Cylinder, even though this description contrasts sharply with the broken condition of the Cylinder when it arrived in London. First, Toma's list does not include another entry that could represent the Cyrus Cylinder. Second, only one cylinder is listed in the receipt inventory compiled by Pinches, an able and incurably inquisitive Assyriologist who was in the habit of reading even the most mundane documents. There are two inconsistencies to explain: the size of the cylinder and its condition. The former is relatively straightforward. Rassam and Toma had until this point been excavating in Assyria, where the 'cylinders' found were much larger objects known to us today as 'prisms'. Other Neo-Babylonian cylinders are also often somewhat larger than Cyrus'. The discrepancy over the condition of the cylinder is harder to explain. One of the earliest known photographs of the Cyrus Cylinder, that published by Rassam in 1897 (*figure 1*), shows clearly the sharpness of the broken edges of the first surviving lines. Thus the object itself suggests that it was probably complete when found and had not been exposed in antiquity.

The obvious question is whether the Cyrus Cylinder could have been accidentally broken in transit between find recording at Babylon and receipt in London. Pinches even comments on the very page where the Cyrus Cylinder is inventoried that,

> with very few exceptions indeed, all the above-mentioned Tablets are of unburnt clay. Many, therefore, have been broken, or otherwise damaged, in transit, by bad packing. Most of the fragments, found as such, are so very small as to be, in their present incomplete state, quite valueless.

The photograph published in Budge (1884) shows that the Cylinder had been glued together from fragments, although whether this was as a result of damage in transit is unclear. And were further fragments present in the cases, Pinches would certainly have found and rejoined them.

A valuable clue is provided by the fact that a sizeable fragment of the Cyrus Cylinder was acquired more than 25 years after these events by the Rev. Dr J.B. Nies in Baghdad, Paris, London, New York or elsewhere. Thus the Cylinder must have been broken prior to its arrival at the Museum. And there is good reason to believe that the missing fragments still await rediscovery among public or private collections.

While it cannot be ruled out that Toma's description of the Cylinder as unbroken was simply incorrect, it does seem improbable. Thus, unless someone substituted the Cyrus Cylinder for another cylinder after Toma's inventory was written, we have to conclude that one or other of the individuals who featured in the transmission of the Cylinder to London was responsible for breaking it and removing one or more parts of it for their own purposes. The theft and sale of antiquities was commonplace at this time. The possible suspects are as follows:

1. The workmen at Babylon. Rassam (1897: 263) tells how dealers had tried to bribe the workers to steal antiquities soon after he had started work. But they may probably be exonerated in this case, since otherwise it would mean that Toma was effectively covering for their misdeed.

2. Toma or Shamas. Both men later sold large quantities of cuneiform tablets to the British Museum and other museums in Europe and America.
3. British and Ottoman officials or their staff, who unpacked, inspected and repacked the antiquities bound for London. We cannot know exactly what happened in Baghdad from our sources, but the process seems to have been that cases were unpacked and inspected with both British and Ottoman officials present.

It thus appears most likely that it was Toma and/or Shamas who would have been responsible.

There is evidence that in the late 1870s cylinders could be broken deliberately for sale. Rassam describes the damage done to another cylinder (identifiable as Rm 673) in two letters to Winter Jones, written not long before the Cyrus Cylinder was discovered:

> I have, however, secured half of a large terracotta Cylinder with fine and well preserved cuneiform characters on it. This it appears has been cut by the Arabs to enable them, as they thought, to get more money for them as they supposed that two small finds would fetch more than one large cylinder! forgetting that they injure the whole by trying to saw it in two. Unfortunately most of the other half is missing as it appears that while they were trying to cut this valuable relic in two they smashed the upper end. (OP 687; 24 December 1877)
>
> … because they smash everything they come upon and on many occasions they broke the tablets for the purpose of increasing their number or for the sake of selling some of the pieces to others. With the purchases I have made for the British Museum I bought a very nice terracotta round cylinder with which most unfortunately the ignorant Arabs had played the same trick; because I found that they had tried to saw it in two in order that they might sell one part to [the local dealers] and the other to somebody else. It seemed that while they were doing this the upper part was smashed into a dozen pieces, and although I managed to find a few of the fragments more than a quarter of the upper part is still missing. (OP 1681; 26 January 1878)

Where was the Cyrus Cylinder found?

The shipment that contained the Cylinder grouped material together as being 'discovered at Telambran and Jumjamah'. Rassam asserted in his letter of November 1879 that the Cylinder was found at Amran, but in *Asshur and the Land of Nimrod* (1897: 267) he says: 'we discovered in the

17 Detail from *Map of the Ruins at Babylon* by W.B. Selby (1859).

ruins of Jimjima a broken terra-cotta cylinder, which has been deciphered in the first instance by Sir Henry Rawlinson, and found to contain the official record of the taking of Babylon by Cyrus…' It is not clear what brought about this change of provenance from Amran to Jumjuma, if indeed it is a change. At that time the site of Babylon comprised a bewildering set of topographical features; Rassam's description of the site (above, p. 44) reveals the scope and nature of the brick mining that had been taking place for centuries. Onto this was mapped a shifting set of toponyms. The nineteenth-century explorers and excavators applied names to parts of the site slightly differently from each other, and on occasion inconsistently. There is no clear description of what part of the site Rassam considered to be covered by Jumjuma, and to a certain extent Amran and Jumjuma seem to have merged into a single area. While the map in Rich (1815) applies the label 'Jumjuma' to a small mound on the remains of the outer walls opposite the village, others attached the name differently. For Budge (1920: vol. I, 288), Jumjuma was the mound on which stood the tomb of Amran ibn Ali. For a variety of reasons it is highly unlikely that Rassam excavated within the village of Jumjuma itself, or within its walled date orchards. There is no trace of the negotiations or compensation that would have been required, for example. Jumjuma perhaps refers to the part of the larger Amran mound opposite the village to the southwest (labelled 'Enclosure'), which was divided off from the main set of mounds by a canal, as seen clearly on W.B. Selby's (1859) map (*figure 17*).[19]

An account of the place of discovery is given in J.P. Peters (1897: 210; based on visits in 1885 and 1889): 'It is the mound of Amran, or Jimjimeh, which is the most fruitful in antiquities; and it is here that almost all of the clay tablets and cylinders have been found which have reached Europe and America'. Here Jumjuma is synonymous with Amran. Likewise Hilprecht (1903: 30) explains that the most southern point of the Amran ruin mounds is connected with the embankment and village, both of which are called Jumjuma, and that 'he [Rassam] concentrated his efforts at the two southern groups of the vast complex, known under the names of 'Omrân ibn 'Alî and Jumjuma' (262). Peters continues (1897: 211): 'On a later visit to Babylon I was shown, by the man who discovered it, the place where the famous cylinder of Cyrus was found. This was on the mound of Amran, but not in the corner of a building. It was in a sort of niche in the face of a long wall.' In such circumstances, one must question how much confidence can be put in the reliability of this claim. The supposed findspot in a wall is plausible enough, however. As discussed above, cylinders were traditionally deposited in niches, and have been found *in situ* and in the remains of fallen walls. Among the construction works described in the Cylinder inscription itself is repair work on Imgur-Enlil (the inner city wall of Babylon) and the quay wall, as well as another wall or building. The inner, outer and quay walls all run near the village of Jumjuma.

The other possibility would be that the Cylinder came from the Esagil, further north.[20] Here we may refer to Rassam's (1897: 349) observations, made in the context of the 1880 season:

> That part of the mound called Omran, to the north of the sanctum of that name, is more mysterious to me than any mound I ever dug at, either in Assyria or Babylonia; because, while the southern portion contained evident signs of ancient remains, where we discovered a large number of inscribed clay tablets, the northern part was an accumulation of ashes, bones, fragments of pottery, and other refuse. We could find no sign of inscription or any object of interest to show that it had ever been occupied.

The Esagil is (a little) north of the shrine, so within the area barren of inscriptions. That would leave the Imgur-Enlil wall as the more likely place of deposition in antiquity, and Jumjuma as the place of excavation. It cannot be ruled out, however, that the Cylinder did indeed come from Esagil, not from its structure, but from what may have been a library building south of the temple.[21] The fact that only a single copy of the Cylinder was found (so early on) in the course of expansive excavations hints that it was found either in a minor repair to an older structure or in an archival context.

The Cylinder's significance realised

The earliest definitive reference to the Cyrus Cylinder comes on 12 September 1879, when Rawlinson wrote excitedly to Birch and Pinches:

> Dear Mr Birch,
> Thanks for your note of the 7th. Pinches has since sent me an account of the Cylinder of Cyrus the Great, containing his genealogy, capture of Babylon &c. This has quite revived my old interest in the subject, and I

am all anxiety to see the Cylinder and investigate the new historical & geographical names. (ME Corr. 5361)

Dear Mr Pinches,
   I am greatly interested in your account of the Babylonian Cylinder of Cyrus, and should much like to have a copy at once of the portion relating to the genealogy, as I know nothing of the city of Ans-an (how is it written?) nor of king Tzras(?), and the notice seems to open into a new field of research into the history of the Far East.
   I expect the summoning by Mr Bond to a special meeting of the Trustees during next week, and could then look over the new tablets, but in the mean time I want to see as much of the cuneiform text of the Cyrus Cylinder as you can conveniently copy out. (ME Corr. 5362).

His first look at the Cylinder itself was not long in coming. By 18 September the Cylinder had already received its modern name:

I am in town for a few days and propose to take advantage of my visit to look at the tablets from Babylon and especially at the Cyrus Cylinder – I shall therefore be at the Museum tomorrow by about 11 o'clock, and if the day is tolerably bright, will look at the Cylinder etc. in the Board room, which is more bright/convenient than Birch's sanctum. (ME Corr. 5514, letter to Pinches)

Rawlinson was growing impatient for Pinches' copy:

If you are short of time and cannot copy the Cyrus Cylinder, please let me know and I will come to the Museum myself tomorrow afternoon and take a copy. I send this as time presses and I am anxious to have the text to analyse and work at. (PC; 4 October)

A letter to Birch reveals his plan to make a spectacular announcement, and fear that other scholars might beat him to it:

I hope you have not put the Cyrus Cylinder into the hands of Oppert or Schrader, as we shall lose the whole credit of the discovery, which I am desirous of announcing at the opening of the Asiat. Soc. Session, early in next month. I believe I have identified Ansan, and can give some other illustrations of interest – but I should like to have any further copies that Pinches may have made of the text. (ME Corr. 5515; 10 October)

By the end of the month he was pressuring Pinches to finish his copy:

will come tomorrow afternoon at about 2. PM to copy Cylinder in the Board Room, unless I hear from you in the interim. (ME Corr. 5516; 28 October)

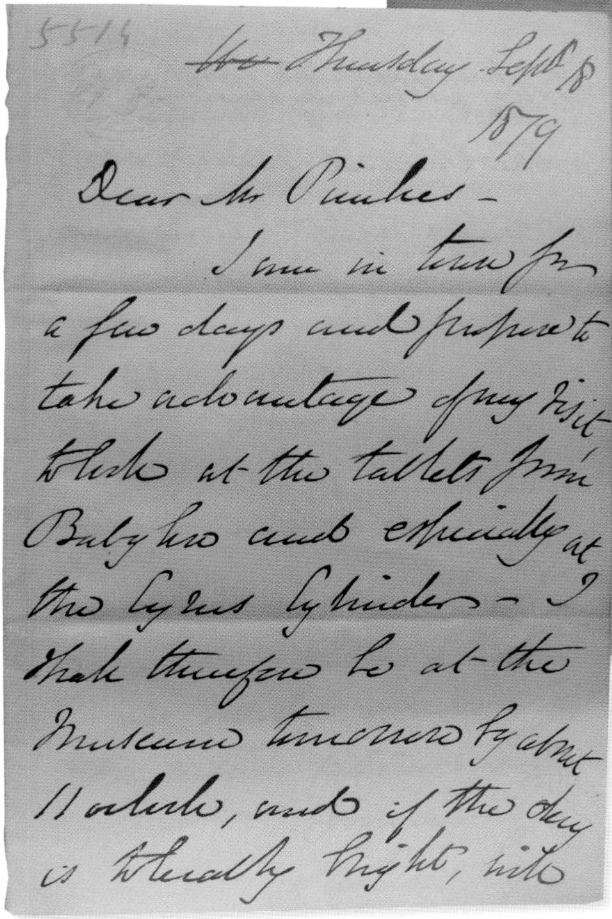

5514

~~He~~ Thursday Sept 18
1879

Dear Mr Pinches —

I am in town for a few days and propose to take advantage of my visit to look at the tablets from Babylon and especially at the Cyrus Cylinder — I shall therefore be at the Museum tomorrow by about 11 oclock, and if the day is tolerably bright, will

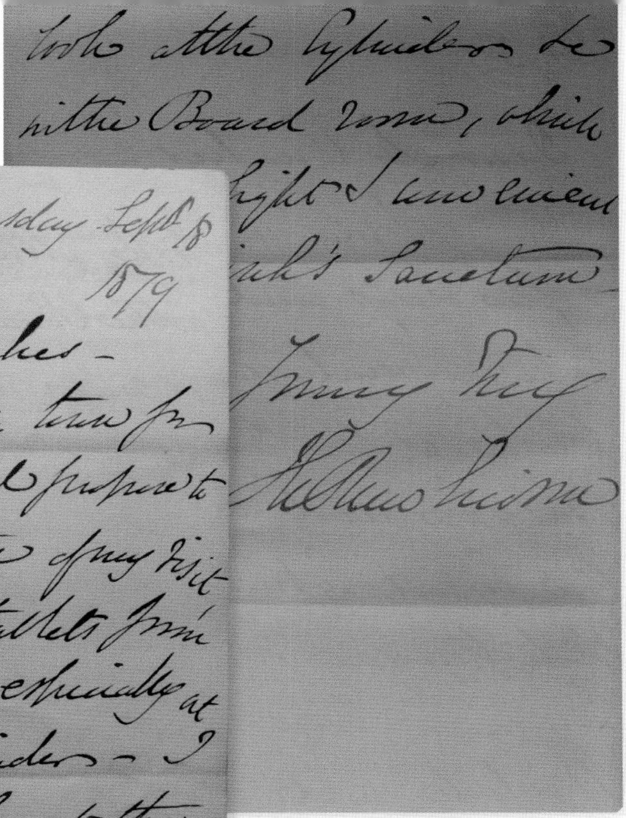

took at the Cylinder in the Board room, which ~~night~~ I am aware is ~~Birch's~~ Sanctum —

Yours truly
H. C. Rawlinson

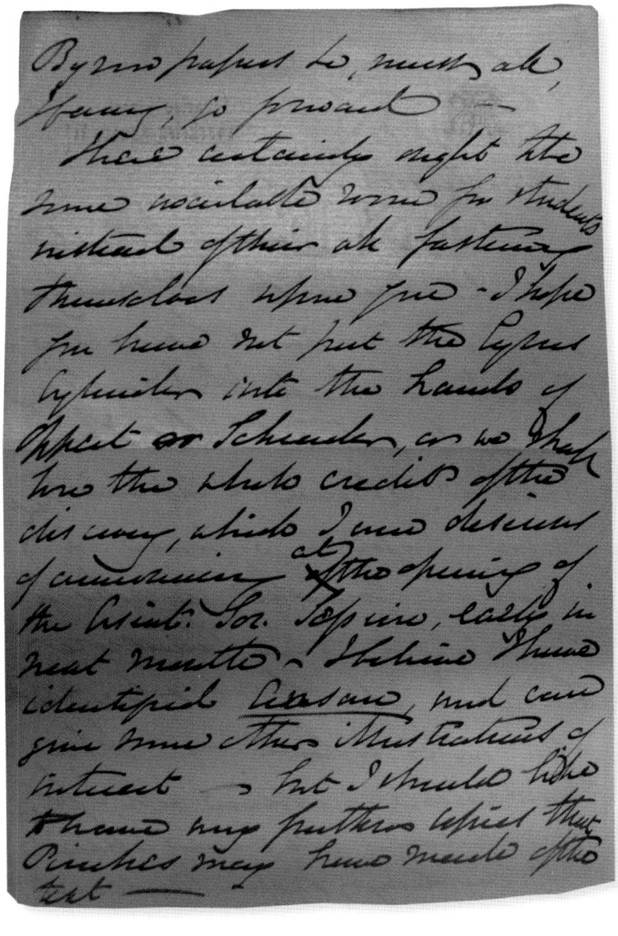

Byrus papers to much at Henry, is present —

There certainly ought to be some available room for students instead of their all fastening themselves upon you — I hope you have not put the Cyrus Cylinder into the hands of Oppert or Schrader, as we shall lose the whole credit of the discovery, which I am desirous of announcing at the opening of the Asiatic Soc. Session, early in next month — I think I have identified Akkad, and can give some other illustrations of interest — but I should like to have any further which Mr Pinches may have made of the text —

18 (*top*) Letter from Rawlinson to Pinches, dated 18 September 1879.

ME Corr. 5514.

19 (*bottom*) Letter from Rawlinson to Birch, dated 10 October 1879.

ME Corr. 5515.

That letter produced results, and perhaps also a suggestion that Pinches may have sabotaged the text:

> I duly received the copy of the Cyrus Cylinder last night – and return you my best thanks for the trouble you have taken – I am surprised to find a few new letters, introduced I presume by Cyrus. (PC; 29 October)

Rawlinson would present his paper at the Royal Asiatic Society just three weeks later. It is clear that credit for identifying the Cylinder should go to Pinches, however. At the time he was working on texts relating to this period of history, and on other historical inscriptions. It was he who was responsible for unpacking the crates of tablets that arrived at the Museum. He compiled the inventory of tablets from Babylon and Borsippa. Pinches identified the Cylinder and began work on it, before ceding to Major Rawlinson. So exciting was the discovery of the Cylinder that it was presented to the public before its arrival at the Museum had been officially reported to the Trustees. The *British Museum Returns to Parliament for 1880* (p. 18) records, among the acquisitions of the previous year: 'Portions of a fine terra-cotta barrel cylinder, containing an account of the taking of Babylon by Cyrus, his genealogy, and entrance into Babylon.' Pinches' copy of the Cylinder would appear soon after as text 35 in the fifth volume of the British Museum's landmark series *The Cuneiform Inscriptions of Western Asia* (1880), described there dispassionately as 'Inscription from a barrel cylinder of Cyrus, from Babylon'.

Summary of key points

For the convenience of the reader, the key moments in the history of the Cyrus Cylinder are summarised here:

1. The Cylinder was excavated at Babylon between 17 and 23 March 1879 by a team under the direction of Daoud Toma. It was found either at Amran (having been placed in the archives of the Esagil) or at Jumjuma (having been placed as a building deposit in a wall).
2. It was probably intact when first discovered, and broken by Toma or Shamas.
3. The main fragment arrived in the British Museum in August 1879.

20 Theophilus Goldridge Pinches, cuneiform curator at the British Museum; the first man to read the Cyrus Cylinder in modern times.

A smaller fragment was later acquired by Nies, becoming part of the Yale Babylonian Collection by 1920.
4. The British Museum fragment was read and understood by Pinches by 18 September 1879, and copied by him in October.
5. It was presented to the public in a talk by Rawlinson in November 1879, and published by him in the *Journal of the Royal Asiatic Society* the following year. The Yale fragment was identified by Berger in 1971.

The Cyrus Cylinder as object; further remarks

Mesopotamian building deposits

The barrel-cylinder shape is characteristic of building deposits of the Neo-Babylonian period. They are the equivalent of the well-known Assyrian prisms bearing annals of the Neo-Assyrian kings. Such deposits fit into a long tradition, stretching back far into the third millennium. This type of deposit is often called 'foundation deposit', since many such objects were buried in the foundations of temples. This was not always the case with the Neo-Babylonian cylinders, however. They could be carefully placed in the structure of important buildings or city walls. The excavators of the Cylinder in 1879 likened its shape to that of a honeydew melon. Examples excavated a year later at Sippar were labelled as 'milestones', referring instead perhaps to their findspots.

It has been suggested (Ellis 1968: 116) that the cylinder may have developed out of the conical deposits popular in the late third and early second millennia. These were tapering clay cones, flattened on one or both ends. Some examples taper from the centre towards both ends, yielding an appearance very similar to that of a later barrel-cylinder. During the first millennium, cylinders are a standard type of Babylonian building deposit; this shape of object could also sometimes be used for scholarly purposes. Cylinders were produced in some numbers, and often several copies of a cylinder deposit have been discovered. This is one reason why we could expect to find further copies of the Cyrus Cylinder in the future.

21 The Cyrus Cylinder and three other Babylonian cylinders: (*top to bottom*) The Cyrus Cylinder; Nebuchadnezzar (from Sippar), BM 91114; Ashurbanipal (from Babylon; it is probably a duplicate of this cylinder that Cyrus in his Cylinder reports having found), BM 86918; Nabopolassar (from Sippar), BM 91105.

The manufacture of cylinders

A variety of Neo-Babylonian 'foundation' cylinder types is attested (see Schaudig 2001: 29–31). The most striking difference is the variation in size, ranging from large examples that measure about 30 cm long to much smaller examples that measure about 7 cm long (*figure* 21). The next most obvious differences are whether the cylinder is solid or hollow, and whether pierced or not. The largest cylinders are hollow. Medium and small cylinders can be either solid or hollow. Deposits of this type would usually be fired for extra longevity. A large cylinder faces a reduced risk of damage during firing when made hollow. It has been suggested that two hollow cylinders of an unusual 'grenade' shape may have contained liquid offerings (Da Riva 2008: 38). The clay used to make cylinders is usually of good quality, with varying levels of temper. Where used, temper is of vegetable matter.

Cylinders were often made on a wheel, as evidenced by the striations on the inner walls of hollow examples. The internal surface of some cylinders shows deep striations where a tool or pebble has been used. Some show deep diagonal striations, suggesting vary hasty manufacture. Cylinders could also be hand rolled instead. Hollow, wheel-made cylinders often have one end thicker than the other. This is an artefact of the manufacturing process. It is also advantageous, since it provides the cylinder with stability when stood on end. And cylinders have been found *in situ*, stood on end in just this manner. Usually the text is written starting at the top end and working down.

Some solid cylinders are pierced along their length, and some hollow ones have holes in the centre of each end. Other cylinders are pierced only at one end, while yet others are not pierced at all. It has been suggested that the piercings are designed to allow rotation on a spindle, to facilitate reading (see Da Riva 2008: 38–9). But while a few examples are known to have been kept in archives or put on display, as a category of object generally they were designed to be placed out of sight and beyond reach. When future generations found them, they would be confronted by objects that are neither particularly heavy nor awkward to handle. Even when present, there is considerable variety and irregularity in the nature of the piercings. This suggests that they are related more to manufacture than intended use.

The production of the Cyrus Cylinder

The Cyrus Cylinder is medium-sized, solid, unpierced and made from tempered clay. In general the physical features of the Cylinder fit well within the range of attested Babylonian practice described above. Some details of its construction, however, are unusual, and consideration of the physical aspects of the Cyrus Cylinder in comparison to other cylinders can help clarify its nature. Most noticeably, the text of the Cylinder is written in a single column. This is not the usual Babylonian arrangement in cylinders of this size, but is typical of cylinders deposited at Babylon and elsewhere during the period of Assyrian rule, a century earlier. This choice is perhaps influenced by the format of the cylinder of Ashurbanipal found during Cyrus' work, as described in the Cyrus Cylinder text.

The choice of temper in the Cylinder's clay is strikingly unusual, maybe even unique. Many stone inclusions can be seen in the clay matrix. They range in size from rather large pebbles at least 5 mm long, down to smaller pebbles about 1 mm long. These inclusions would have facilitated an even distribution of heat through the object as it was fired. This is particularly important when firing objects as thick as a solid cylinder. It is possible that these pebbles were already in the raw clay chosen to make the Cylinder, but it is nevertheless significant that they were not removed. Photographs taken at the time of the replacement of the conservation fill at one end of the Cylinder suggest that an outer layer of clay had started to come away from the core (see *figure 5*). This weakness in the body would influence how it later broke. The depth of the Yale fragment matches that of the outer section.

The Cylinder displays a range of colours, from light beige to deeper browns. This is a product of the firing process. BM 90920 was re-fired in 1961, but the loan of the fragment from the Yale Babylonian Collection, NBC 2504 was agreed only in 1972. The colorations are shared across the two pieces, and thus must be a product of uneven firing in antiquity. This unevenness suggests something about how the kiln was stacked. The Cylinder would have been fired together with similar objects, quite probably duplicates of the same inscription. This provides further reason to be optimistic that future excavations may yield duplicates of the Cylinder, restoring the missing portions of text.

NOTES

1. *The Times*, Tuesday 18 November 1879.
2. An area of the site of Babylon also referred to as Amran (ibn Ali).
3. Other material which arrived in the same shipment as the Cylinder was known to have come from Birs Nimrud; the Cylinder itself, however, was known to have come from Babylon.
4. A formal permission from the Ottoman ruler, in this case to undertake archaeological excavations.
5. The Kasr (or 'castle'), another area of the site of Babylon.
6. Rassam to the principal librarian, from Mosul, 21 March 1878: 'It is desirable to continue explorations on large scale in April and send agents to Babylon to dig for inscribed tablets will trustees consent' (OP 1788).
7. The modern town near the site of Babylon.
8. By this he means October (OP 1480; 12 January 1878).
9. The Islamic historian al-Qazvini records this practice already in the thirteenth century. Babylon supplied bricks for towns up and down the Euphrates.
10. It is ironic that Budge, as keeper of the Department of Egyptian and Assyrian Antiquities, would later slander Rassam with charges that he and his family had been behind the looting. Layard, by contrast, described him as: 'one of the honestest and most straightforward fellows I ever knew' (quoted in Waterfield 1963: 483).
11. He refers here to Baghdadi antiquities dealers paying locals to dig for antiquities.
12. Rassam (1897: 285) explains how the shipment was delayed by a week when the Ottoman inspector struggled to understand the contents and required a second inspection.
13. Rassam, in the context of Assyrian excavations, explained his procedure in detail:

    Generally speaking, my workmen excavated by gangs of seven, – a digger, a basket-filler, and five basket carriers, – that is to say, those men who carried away the débris from the trenches. But on certain occasions, when the rubbish had to be carried far away, the basket-carriers used to be augmented from those gangs who had a shorter distance to dispose of their load. In each separate mound I generally placed Christian overseers, because they knew how to read and write; and if the work became extensive I placed under them one or two Arab sub-overseers. (1897: 199)

    For his excavations at Babylon he managed to negotiate a favourable rate with the workers, settling at a level one-third lower than normal (Rassam 1881: 211).
14. For instance, see Finkel 2008: 171; Reade 1986: xix; Walker 1972.
15. His own account of this period can be found in Rassam (1897: esp. pp. 258–91).
16. The Arabic actually says 'true "honeydew melon" [*shamama*], unbroken, small'. Here the melon is a reference to the shape of barrel cylinders. My thanks to Lamia al-Gailani Werr and Venetia Porter for assistance with the translation and interpretation of this passage.
17. Consul Miles reports (OP 3984; 20 August, letter to the principal librarian) that the cases which arrived during his tenure were opened in his presence and that of the Ottoman representative, without anything being removed. Nixon ought to have followed the same procedure.
18. This is the same batch of tablets that Rassam mentioned in his 20 November letter. The discrepancy between this figure and the 600 and 800 tablets mentioned in that letter is presumably due to the heightened emotion of the moment.
19. In October/November 1879 Toma was still excavating the mounds opposite Jumjuma village, near to the remains of the city walls (see Reade 1986).
20. The sanctuary of Babylon's tutelary god, Marduk; for this suggestion and references to such inscriptions being located in archives see Schaudig 2001: 46.
21. For the Esangil archives, see Clancier 2009, esp. pp. 122–3, 168–81, 203–5.

3

The Cyrus Cylinder: display and replica

ST JOHN SIMPSON

The display of the original

THE Cyrus Cylinder has been more or less continually displayed in the British Museum since its discovery, transport and unpacking in the British Museum in 1879. The style of display has changed according to the gallery or case in which it was exhibited. It was originally displayed as part of the Babylonian collection in a table-case of other cuneiform inscriptions relating to 'the great building operations carried out in Babylon and other cities by kings of the last Babylonian Empire' (British Museum 1922: 137) (*figures 22–24*).

The situation changed in 1931 when the cylinder was exhibited in a major Persian Art exhibition at the British Museum. This was held from January to May in the Sub-department of Oriental Prints and Drawings in the upper galleries of the King Edward VII Building and brought together objects from seven departments: Egyptian and Assyrian Antiquities (then a single department), Greek and Roman Antiquities, Oriental Manuscripts (now part of the British Library), Prints and Drawings, Ceramics and Ethnography (then including the Islamic collections), British and Medieval Antiquities (the forerunner of the present Department of Prehistory and Europe, which still held the Oxus Treasure as part of the Franks Bequest), and Coins and Medals. The accompanying printed leaflet describes how 'the "Cyrus Cylinder" bears a long inscription celebrating the conquests and

[12049] PORTION OF A BAKED CLAY CYLINDER OF CYRUS, KING OF BABYLONIA, ABOUT B.C. 538—529, INSCRIBED IN THE BABYLONIAN CHARACTER WITH AN ACCOUNT OF HIS CONQUEST OF BABYLONIA, AND OF THE CHIEF EVENTS OF HIS REIGN IN THAT COUNTRY. CYRUS ATTRIBUTES THE SUCCESS

piety of this king. It reveals that his capture of Babylon was bloodless and probably obtained by treachery' (British Museum 1931: 6). This huge exhibition was triggered by the simultaneous blockbuster exhibition on *Persian Art* which opened at Burlington House on 7 January (Wood 2000).

Following the success of the exhibitions and the wave of interest in Iran, the new 'Persian Room' was opened at the British Museum later that same year. This was the first dedicated permanent display on ancient Iran in the upper galleries and replaced the earlier ground floor displays which had been simply based on sculptures and plaster casts of Persepolitan reliefs. This was a small gallery located next to the North-East Staircase which had been previously used to display antiquities from Coptic Egypt and which corresponds to the present display of Early Anatolia in Room 54 (British Museum 1932: 18, 32). This gallery, then numbered Room 20, was rearranged by Sidney Smith to display those 'antiquities from Persia' which had been temporarily shown in the *Persian Art* exhibition and/or previously exhibited in the Babylonian and Assyrian galleries, and included the Oxus Treasure and assorted Achaemenid and Sasanian silverwares which were specially transferred from the Franks Room display of the British and Medieval Department. The report on what was described as the Museum's first 'exhibition of antiquities from Persia and Armenia' specifically described how the 'gold objects of the Oxus Treasure have been remounted on a new fitting and exhibition arranged' (*Reports to the Trustees*, 6 August 1932; 5 January 1933). Meanwhile, the Cyrus Cylinder was displayed in the centre of a wall-case crammed with inscribed alabaster vase fragments, bricks, tablets, weights and part of a reconstructed miniature plaster column (*figures 25–26*).[1] It was therefore primarily used as an example of a royal inscription within the context of the Persian Empire. This display was dismantled in 1936 in preparation for World War II when all of the Museum's galleries were either emptied or sandbagged (Caygill 1981: 53–5).

After World War II the museum's collections were returned from off-site storage and the displays were gradually restored in the old wall-cases and table-cases. By 1952 the Persian Room display had been restored to the western half of the spot later occupied by the Room of Writing and present-day Later Mesopotamia gallery (*figure 27*). The 1952 guide to the Western

22 (*top*) Old gallery photograph of the Cyrus Cylinder with its gold-edged display label.

23 (*centre*) Another view of the Cyrus Cylinder on its original mount.

24 (*bottom*) A nineteenth-century view of the Cyrus Cylinder.

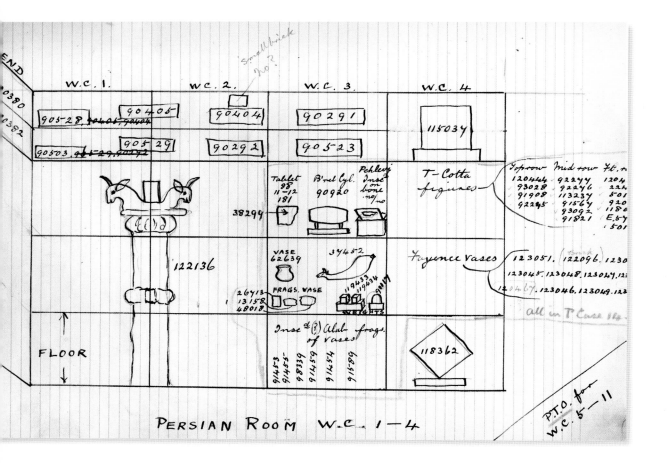

25 Sketch elevation showing the position of the Cyrus Cylinder (90920) in wall-case 3 in the Persian Room, drawn prior to the dismantling of the display in 1936.

Asiatic displays makes no mention of the Cyrus Cylinder. It highlights instead the 'pre-historic painted pottery from various sites in West and North Iran [cases 1–2, 17–18] … the Oxus Treasure, gold and silver ornaments and vessels of exquisite workmanship from the Achaemenid period (5th century BC) down to the Sasanian (3rd–7th centuries AD) … [and] the Luristan bronzes from West Iran; horse-furniture, vessels, weapons, and tools, mostly about 1200 BC [wall-case 4, centre-case C]' (British Museum 1952: 14). Plans and case drawings of this display indicate that the objects were arranged either on existing shelves in the old wall-cases or in a small number of table-cases. It is for this reason that the displays were relatively easy and cheap to mount or change, and, unlike today, such refurbishments were carried out entirely by the curatorial departments concerned. This

26 Part of a plaster model of an Achaemenid column capital from Susa, formerly displayed in the Persian Room.

BM C[ast] 34.

was not without some external criticism, though, as the displays were attacked either as over-academic (Hawkes 1962) or jumbled (Anon. 1962). The drawings also show the Cyrus Cylinder placed in the centre of the top shelf in a wall-case (case 15), together with fragmentary stone vases and weights with royal Achaemenid inscriptions, thus resembling the pre-war display in concept and design (*figure 28*). This room was later renamed the Iranian Room, before it was moved by Barnett in 1958 to the top of the West Stairs as the so-called Persian Landing where it replaced the former Hittite Landing (*figure 29*).[2] The new location was considerably smaller than the previous gallery and the display had to be heavily reduced as a result, yet the position at the top of the main West Stairs was a prominent one on a popular circulation route for the public.

In the same year (1958), planning began in Tehran for the Celebration of the 2,500th Anniversary of the Founding of the Persian Empire by Cyrus the Great and a Central Committee was created for that purpose. These celebrations were originally scheduled for 1962, thus 2,500 years after Cyrus' entry into Babylon, but were later postponed. The Shah's coronation celebrations on his 47th birthday on 26 October 1967 provided a foretaste of what was eventually to come. One witness to these events was

DISPLAY AND REPLICA

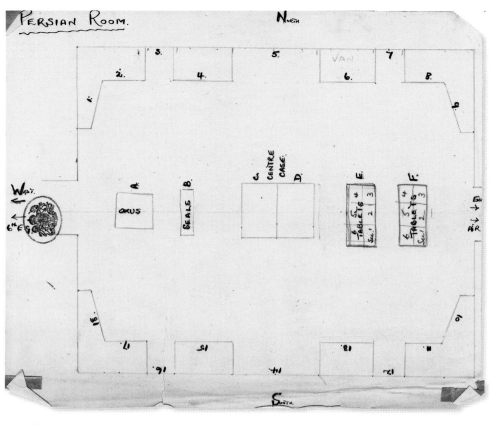
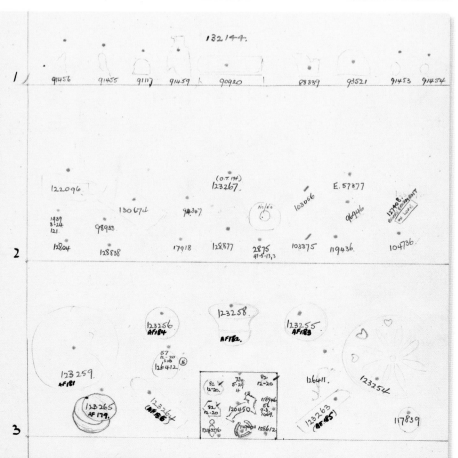

27 (*left, top*) Sketch plan of the Persian Room.

28 (*left, bottom*) Sketch elevation showing the position of the Cyrus Cylinder (90920) in case 15 in the post-World War II Persian Room, 1950s.

29 (*right*) Plan of the British Museum galleries indicating the position of the Persian Landing, 1961.

the Reverend Norman Sharp (1896–1995) (*figure* 30).³ Writing in August that year (1967) to his friend and Shiraz contemporary, the late Paul Gotch (former director of the British Council regional centre in Shiraz from 1959 to 1966), Sharp gives some very interesting insights into this period:

> The coronation takes place on the Shah's 47th birthday, October 26th, and this is being separated from the Cyrus celebrations. The University is bringing out a special publication in connection with the coronation, and I have been asked to write an article for it, which will contain quotations in cuneiform.... Two weeks after the coronation the Shah is to open a new port in the Persian Gulf, called Mashahr. The Refinery authorities wish to make the Shah a presentation on that occasion. They decided to have a beautiful gold bowl made by goldsmiths in London, bearing Achaemenian designs, resting on the back of four gold gazelles, such as are sometimes to be seen in the desert near this port, and they will be standing on a block of onyx. (I begged them to change the onyx for one of the beautiful alabasters or marbles of Persia, but nothing came of this). They desired to have an inscription engraved on the bowl in cuneiform, and told [Dr Assadullah] Alam [the Chancellor of Pahlavi University] they thought of approaching the Teheran University about it. He told them that they should apply to the Pahlavi University,⁴ so this is something else that has fallen to me, and the dedication is to be in cuneiform round the bowl. I have set up the type of it: it will be the first time London Goldsmiths have engraved Persian cuneiform. It will somewhat resemble the beautiful gold bowl of Xerxes in the Archaeological Museum.⁵

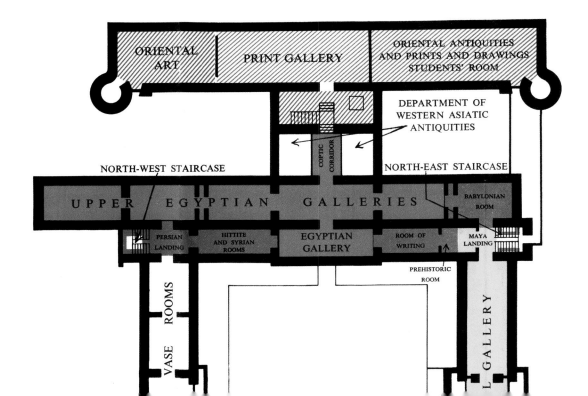

The celebration of the Cyrus initiation of the Persian Empire is to be put forward four years. The coronation is more or less an occasion for Persians: the other celebration will be on a much larger scale … I think I mentioned it was proposed to put up in Teheran a great block of stone in a prominent place, on one side of which on a great sheet of unrusting metal in raised letters would be the Shah's new charter on land reform etc, and on the other side Cyrus' charter when he entered Babylon. I mentioned it to Bushahri that Cyrus of course would not be able to read it, as he used another script, and that at least his name and title should be in Persian cuneiform. Bushahri said he thought it was not necessary, and that there would not be room on the sheet of metal, so I wrote to the Council on the subject. This was discussed and decided on some occasion when Bushahri was absent, and one day when I was standing in Bushahri's room two letters were brought in from the Council, one for him and one for me, stating that the cuneiform words were to be attached in raised letters to the plaque. Then when I was in Bushahri's house I was surprised to see another copy just like the first. They then told me that 160 were to be made, one for each of the 160 cities, towns and ports of Iran, so eventually in all of these places, the name of Cyrus in the original cuneiform will be seen, but no one will know what lies behind this achievement.[6]

30 Reverend Norman Sharp (*left*) with Ali Sami at Persepolis (after Sami 1970: 176).

The Cyrus Cylinder was loaned to Tehran for a brief period of exhibition from 7 to 19 October 1971 at the specially constructed Shahyad (now Azadi) Monument in Tehran (Bailey 2004). It was taken out and subsequently collected personally by Barnett, and ten days later it featured in a hastily constructed temporary exhibition on *Royal Persia: A Commemoration of Cyrus the Great and His Successors*, which opened at the British Museum on 29 October (Pinder-Wilson 1971) (*figure 31*). This was jointly opened by Sir John Wolfenden, the chair of the Museum's Trustees, and His Excellency Amir Khosrow Afshar, the Iranian ambassador in London, who later made an official request for a cast (see below).[7] This display remained until 30 January 1972 and replaced the previously programmed exhibition in the King's Library on *Cooking in the Orient*. It was during this period that the former Persian Room was dismantled and, following

31 Cover of the catalogue of the British Museum exhibition *Royal Persia* (1971).

the planned departure of the ethnography collections to the Museum of Mankind and the vacated space, allocated for a new gallery for Ancient Iran, which opened on 23 July 1975 (*Report of the Trustees 1975–1978*: 39). This was the first of this succession of Ancient Iran galleries to employ a low suspended ceiling, artificial lighting and a series of low but deep new wall-cases arranged end-to-end around the walls, but the object displays were fundamentally very similar to their predecessors (*figures* 32–33).

This Ancient Iran gallery was dismantled in the early 1990s in preparation for a more fundamental structural redevelopment of the galleries along the entire upper east range of the Museum. A temporary display was mounted in 1994 at the southern end while construction was carried out at the opposite end (*figure* 34), followed by a more permanent display at the northern end in 1995 (*figure* 35). Both reused the cases from the 1975 display in order to reduce costs, although a limited budget was released for the 1995 display as it was recognised that the gallery lifespan had to outlast the opening of the Great Court in 2000 and be part of the radical new public circulation routes facilitated by the redevelopment of newly vacated British Library spaces. As with all other displays since 1931, the Cyrus Cylinder was heavily contextualised within the historical framework of Achaemenid royal inscriptions and monumental building. This finally changed in 2003 when it came to be displayed as a single object at the finale of the exhibition on

the Achaemenids: *Forgotten Empire* (Curtis and Tallis 2005). This principle has been re-adopted in its most recent gallery setting at the British Museum, the Rahim Irvani gallery for Ancient Iran, which opened on 21 June 2007. In this case the requirement was to place the object central to the gallery on the main axis of visitor flow, thus enabling it to be seen in the round but therefore necessitating new cosmetic conservation of the broken areas which had not previously been exhibited (*figure* 36).

The Cyrus Cylinder has thus been almost continuously exhibited at the British

DISPLAY AND REPLICA 77

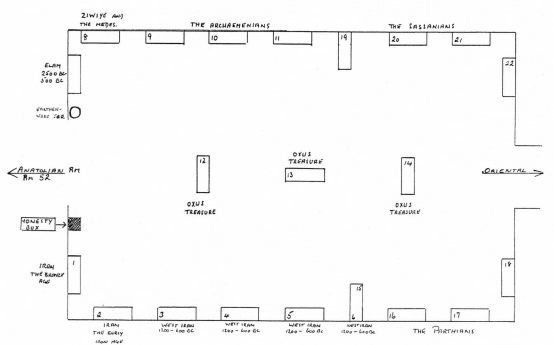

Room 51 Western Asiatic Antiquities.

The Iranian Room

32 (*left*) View and sketch plan of the Iranian Room in 1975.

33 (*right, top*) View of the Cyrus Cylinder case display in the 1975 Iranian Room gallery.

34 (*right, middle*) View of the Cyrus Cylinder case display in the temporary 1994 Iranian Room gallery.

35 (*right, bottom*) View of the Cyrus Cylinder case display in the 1995 Iranian Room gallery.

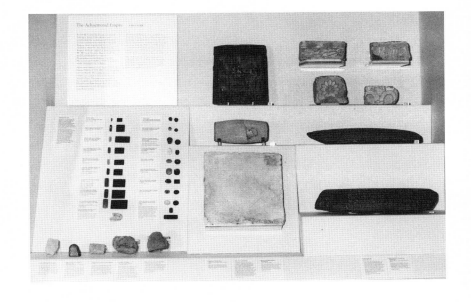

Museum since the late nineteenth century. It has been loaned twice to Tehran and forms the centrepiece of an exhibition touring across America in 2013. The increasing level of international interest in this object, particularly since 1958, occasioned the manufacture of replicas and facsimiles. The story behind these is briefly explored below.

The cast

The original cylinder, as was the normal practice, was fired before burial in the deposit at Babylon. In 1961 the Cyrus Cylinder was re-fired in the British Museum as part of a wider programme of conservation treatment within the tablet collections. Although this was unusual for a previously fired object, it may reflect the need to consolidate the cylinder prior to moulding. It was moulded for the first time in the following year in response to a formal request for a plaster cast from the minister of the imperial court of the Shah of Iran in preparation for the 2,500 jubilee originally planned that year (although it was not listed as commercially available: cf. British Museum 1963). This was followed by a second (personal) request

36 View of the Cyrus Cylinder case display in the 2007 Rahim Irvani Gallery for Ancient Iran.

37 (*right*) Replica of the Cyrus Cylinder in its presentation box.

38 (*left*) The crest on the top of the box containing the replica.

by Norman Sharp (see above) for a plaster cast in August 1971, following an earlier request for an up-to-date translation.⁸ This cast was made from the original mould and taken to Persepolis by Sharp, who presented it to Ali Sami of the Persepolis Museum. A photograph purporting to be of the original object placed on top of a column base at Persepolis is published by R.M. Ghias-Abadi (2001: 20–21); this is certainly not the original object but it could well be this cast. A third request was made by the Iranian ambassador at the opening reception of the British Museum exhibition and relayed in writing by Mr Shapurian, the press attaché of the Iranian embassy in London, who was informed by Ralph Pinder-Wilson

that it 'would take about two weeks to make' using the original mould (letter dated 2 November 1971). On 14 October 1971 one of these casts was presented to the UN secretary general, Sithu U Thant, on behalf of the Iranian government by Princess Ashraf Pahlavi, the Shah's sister and chair of the delegation of Iran to the General Assembly.[9] This cast was placed on display in a showcase designed by the Metropolitan Museum of Art in the corridor outside the Economic and Social Council Chamber along with translations into English and French, and remains on display there today.

Secondary casts remoulded from one of the British Museum casts ordered by Tehran were distributed by the Shah of Iran, and one was presented to the ruler of Umm al-Qaiwain and is currently displayed in the local fort museum. Another was presented to Barnett himself when he took the original to Tehran: this is a pink coloured plaster cast mounted in a red velvet presentation box (*figures 37–38*).[10]

Such casts have also been sold commercially by the British Museum, the National Museum in Tehran and various commercial companies since that period. Those sold by the British Museum were marketed in a series entitled 'Biblical Archaeology' (British Museum 1992). A modified cast was made after the join of the Yale fragment when the object was sent for moulding a second time by Mr Prescott between 7 May and 13 August 1975. The Museum has records of objects moulded for plaster casts, and the Cylinder was one of only 17 Babylonian prisms and tablets reproduced between 1956 and 1975, ranging from the Flood tablet and several Amarna Letters to the Taylor Prism, the Persian Verse Account of Nabonidus and Nebuchadnezzar's first campaign against Jerusalem. Interestingly, the Cyrus Cylinder had not been considered sufficiently significant to be moulded and cast in the nineteenth century, when the focus for replicas was rather on Assyrian or older objects and reliefs.

The changing context of display and rise of interest in replication of the original Cyrus Cylinder closely reflect its transformation in popular public perception from a Babylonian historical document into an icon of Iranian identity. The story of how it has been loaned twice to Tehran and how it is viewed from an Iranian perspective is explored in the following two chapters.

NOTES

1. This column had been modelled on remains from Susa and salvaged from the Persian Court Dining Room, also known as 'Sutton's Folly', at 47 Brook Street in Mayfair, which had been demolished shortly before to make way for Claridge's. This had been completed in 1905 in the house of Sir John Bland Sutton (b. 1855) and was some 12 feet high with a glass roof supported on 32 columns, decorated with turquoise glazed bricks showing Susian archers, and with white linen and purple curtains inspired by descriptions in the Book of Ezra with representations of Artaxerxes based on the Achaemenid Archer Coin Series. Sutton's silver dining set is also inspired by Achaemenid designs; it is currently in the Hunterian Museum of the Royal College of Surgeons.
2. Barnett supplemented the display in 1965 with photographic views of Persepolis: the photographs themselves were taken especially by the late Paul Gotch (1915–2008). This display was dismantled and replaced in February 1976 by the so-called South Arabian Landing after the Iranian collection had been transferred to a new display (*Report of the Trustees 1975–1978*: 39).
3. Sharp lived in Iran for 43 years, following his arrival in 1924 as an Anglican missionary for the Church Missionary Society in Iran. In this capacity he founded churches in Yazd (1928), Shīrāz (1938), Qalat (1944), Bushehr (1944) and Kirmān. Sharp's most famous work was the Church of St Simon the Zealot (Kelīsā-yé Moqaddas-é Sham'ūn-é Ghayūr) in Shīrāz. He designed a seven-line inscription in Old Persian cuneiform for this church, which was cut from stone taken from the original Achaemenid quarry near Sivand (which had only been rediscovered in 1954); it was unveiled in 1971 on the occasion of the Shah's commemoration of the founding of the Persian Empire by Cyrus. The inscription reads: 'God chose Cyrus, and made him king in this earth. May this land of Cyrus be always happy! Honoured be the good name of Cyrus.' Sharp's time in Iran was not always easy, however, and he was put under house arrest during the Mossadeq period, narrowly avoiding expulsion in 1953. In 1954, following the sudden death of Shiraz University's lecturer in Middle Persian, Mr Nikola Rāst, 'a white Russian, who escaped into Iran at the time of the Bolshevik Revolution, and took Persian nationality, and was appointed Head of the Shīrāz Customs Office', Sharp was invited by the head of the Faculty of Letters, Dr Suratgar, to fill the vacancy. He agreed to do this 'provided I could teach the Old Persian language with the cuneiform character, as well as the Pahlavi, which succeeded it in Sassanian times, for there are in Persepolis and elsewhere in Iran many inscriptions deeply cut in stone and well-preserved in Old Persian cuneiform which no one could read, except those who had been taught the script and the grammar of the earliest form of the Persian language. This was welcomed by the students, who after studying the text of an inscription in class, then went to Persepolis, and were able to read the inscription on the stone, and translate it into modern Persian. Some of them became quite adept' (BM Archive/Department of the Middle East/Gotch Papers, letter from Sharp to Dr Ghorban, 17 January 1984). In 1962 Sharp retired from his Church position and became Assistant Professor in the University, where he remained in post until his reluctant retirement in 1967. He also created what remains a very useful introduction to Old Persian, with copies, transliterations and translations of the major monumental inscriptions (Sharp 1966; translated into Persian in 1971 and later reprinted on several occasions). This had been a major undertaking and involved obtaining a copy of the cuneiform font from Germany with the help of his friend Hussein Ala, the Court Minister. Less well known is the fact that Sharp provided David Stronach with the first decipherment of one of the stone copies of the Daivā inscription of Xerxes, which Stronach had just found as a drain cover on the Tall-i Takht at Pasargadae: this followed an urgent

request by the excavator, hopeful that it 'is an early foundation tablet of Cyrus the Great; in any event it must contain a mass of historical information' (Gotch Papers: undated letter [1961]; cf. Stronach 1965: 19–20 n54, pl. V; 1978: 152, pls 122b–123, 161b). According to Gotch's address at Sharp's memorial service, Sharp translated it over lunch at the site. He also translated Persian poetry into English and a book by Sayyed Muhammad Taqi Mostafavi, the former director of the National Museum in Tehran, entitled *The Land of Pars* (1978).
4. Shiraz University was renamed this in 1962 following a visit that year by the Shah.
5. This famous bowl has been regularly illustrated and exhibited as an example of Achaemenid precious metal, although its authenticity is questionable (e.g. Curtis and Tallis 2005: 112, cat. 97).
6. This letter is part of the Gotch archive in the Department of the Middle East at the British Museum.
7. The ambassador's speech omitted reference to the Cyrus Cylinder but he instead dwelt on the role of Britain in 'discovering and understanding the history of Iran', highlighted the role of Sir Henry Rawlinson – 'by his determination, intellectual brilliance and physical endurance, he was the first to decipher completely the Babylonian and Old Persian cuneiform' – and commended the British Museum 'as an institution which has made and is making a unique contribution to the study of my country's past. This it achieves in two ways. First its wonderful collections of antiquities and written documents provide scholars with the means of deeper research into the past and, secondly, it reveals to an ever wider public the history of human achievement' (BM Archive/Department of the Middle East/Royal Persia 1972).
8. Letter of 23 June 1971, addressed to Mr P. Waley in the Department of Printed Books and Manuscripts (British Library), forwarded to Mr C.B.F. Walker in the Department of Western Asiatic Antiquities, who replied enclosing a copy from Pritchard (1950: 315–16) and adding that the object was on display in the Persian Landing. Sharp replied on 31 August that he was still awaiting a reply from the Cast Service but that 'I would much like, if possible to take a cast with me, when I visit Iran for the celebrations.'
9. United Nations Press Release HQ/264, 14 October 1971.
10. Registered as British Museum C.209.

4

The Cyrus Cylinder: the creation of an icon and its loan to Tehran

JOHN CURTIS

THE significance of the Cyrus Cylinder as an important historical document was realised immediately after its discovery by Theophilus G. Pinches, who identified it in 1879 and by Sir Henry Creswicke Rawlinson who published the text in 1880 (see Chapter 2). However, there was no thought at that time that the Cylinder had special connotations for human rights or freedom of expression and there was no hint of the notoriety that the Cylinder was later to acquire.

In Hormuzd Rassam's account of his excavations at Babylon and elsewhere, eventually published in 1897, he describes the Cyrus Cylinder as 'a broken terra-cotta cylinder, which has been deciphered in the first instance by Sir Henry Rawlinson, and found to contain the official record of the taking of Babylon by Cyrus' (Rassam 1897: 267). In the third edition of the British Museum's *Guide to the Babylonian and Assyrian Antiquities* (1922: 144) the Cylinder is described rather dryly as a 'portion of a baked clay cylinder of Cyrus, king of Babylon, about 538–529 BC, inscribed in the Babylonian character with an account of his conquest of Babylon, and with the chief events of his reign in that country'. In the first edition of the *Cambridge Ancient History* (1926), the chapter on 'The foundation and extension of the Persian empire', by G. Buchanan Gray, only refers to 'the Cylinder Inscription', and that in a footnote (on p. 13). In the fourteenth edition of the *Encyclopedia Britannica* (1929), the article on Cyrus the

Great simply refers to 'the cylinder containing his proclamation to the Babylonians'.

Thus matters stood until the 1960s, when the phrases 'first bill of human rights' or 'first declaration of human rights' seem first to have been coined, but exactly by whom is not quite clear. In his 1961 book, *Mission for my Country*, Mohammad Reza Shah Pahlavi makes a number of adulatory references to Cyrus, but he says nothing specific about the Cylinder. For example: 'It was characteristic of Cyrus the Great that, when he conquered Babylon, he allowed the Jews, who had been exiled there by king Nebuchadnezzar after the conquest of Jerusalem in 597 BC, to return to Palestine', and 'Whenever Cyrus the Great conquered, he would pardon the very

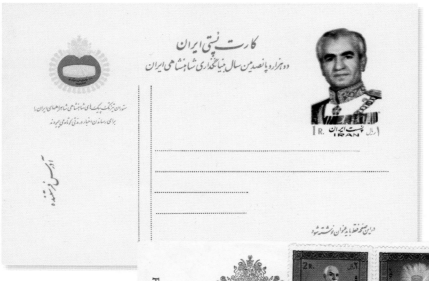

39 (*top*) Prepaid postcard with image of Mohammad Reza Shah Pahlavi and logo of the 2,500th anniversary of the founding of the Persian Empire featuring the Cyrus Cylinder.
Private collection.

40 (*bottom*) First-day cover with set of stamps issued on 12 October 1971 to commemorate the 2,500th anniversary of the founding of the Persian Empire by Cyrus the Great.
Private collection.

41 Stamps featuring the Cyrus Cylinder issued by Iran, Ethiopia and Romania.

Private collection.

people who had fought him, treat them well' (Pahlavi 1961). However, in *The White Revolution of Iran*, published in 1967, the Shah *does* refer to the Cyrus Cylinder, calling it 'the Charter of Liberty granted by Cyrus the Great' (Pahlavi 1967). This claim was repeated in 1968 when he opened the first United Nations Conference on Human Rights in Tehran and referred to the Cyrus Cylinder as 'the precursor to the modern Universal Declaration of Human Rights'.

Meanwhile, the idea of celebrating the '2500th Anniversary of the Founding of the Persian Empire by Cyrus the Great', to give it its official title, had been formally proposed in 1958 (SH 1337) by Dr Shojaeddin Shafa, a man of letters, historian, journalist and Iranian diplomat who was also cultural adviser to the Persian court.[1] Others at the time who apparently enthusiastically supported the idea were André Malraux, the French minister of culture under de Gaulle, and David Ben-Gurion,[2] prime minister of Israel 1948–54, who was an ardent admirer of Cyrus on account of his favourable treatment of the Jews.[3] Shafa gave a number of interviews at this time putting forward the idea, and in due course he secured the formal agreement of the Shah. Initially the celebration was planned for SH 1340 (1961) – that is, 2,500 years after Cyrus' capture of Babylon in 539 BC – but for reasons that are now obscure it was postponed twice.

Eventually, the 2,500 year celebration of the Persian Empire took place on 12–16 October 1971. A large 'tent city' was built near Persepolis to accommodate the 600 guests from around the world, including members of royal families and a large number of presidents and prime ministers. Britain was represented by the Duke of Edinburgh and Princess Anne, the USA by Vice President Spiro Agnew, and France by Prime Minister Jacques Chaban-Delmas. The grand gala dinner on 14 October was one

CREATION OF AN ICON

of the most extravagant meals ever staged, with the catering and food being provided by Maxim's of Paris. On the next day (15 October) there were military parades representing different periods of Persian history, with soldiers dressed in period costume. On the last day (16 October) the Shah inaugurated the Shahyad Tower (renamed the Azadi Tower after the Iranian Revolution), an enormous arch-like monument in Tehran built to commemorate the 2,500-year celebrations. In the base of this tower was a museum where the Cyrus Cylinder was briefly on display until 19 October. It had been taken to Tehran by Dr Richard Barnett, then keeper of the Department of Western Asiatic Antiquities at the British Museum, who was attending the International Congress of Iranologists at Shiraz (13–16 October 1971).[4] The curious story of this loan has been fully researched and published by Martin Bailey in the *Art Newspaper* for 1 September 2004 (see also Chapter 3). As Martin Bailey has deftly shown, 'the Shah used the presence of the Cyrus Cylinder to argue that Persia had been the birthplace of human rights', although the Shah's own record in this respect fell somewhat short of the ideals thought to be embodied in the Cylinder. Inevitably there were calls for the loan to be extended, but these Barnett was able to deflect by pointing out that that the Cylinder was going to be included in the British Museum's own exhibition of *Royal Persia* that was scheduled to open on 29 October.

The Cylinder itself became the official symbol of the celebrations. A logo was produced showing the Cylinder surrounded by a garland of lotus flowers in Persepolitan style and surmounted by the imperial coat of arms of Iran (*figures* 39–40). This features lions with scimitars on either side of a shield with the Pahlavi crown at the top. The Persian inscription beneath the Cylinder reads '2,500th Anniversary of the Founding of the Persian Empire'. In Iran itself a set of four stamps commemorating the Anniversary was issued on 12 October 1971 featuring Mohammad Reza Pahlavi, his father Reza Shah Pahlavi, the Pahlavi crown, and the Cyrus Cylinder (*figures* 40–41; Farahbakhsh 2010). The Cyrus Cylinder also appeared on gold, silver and bronze medals (*figures* 42–3) and on coins (Curtis 2011). A special silver proof set of coins was issued in a presentation case, with the 75 rial coin showing the Cyrus Cylinder (*figure* 43). Surprisingly, neither the Cyrus Cylinder nor the Tomb of Cyrus at Pasargadae that had first

42 (*top*) Bronze medal commemorating the 2,500th anniversary of the founding of the Persian Empire presented to Dr R.D. Barnett.

British Museum 1972,0614.1.

43 (*bottom*) Iranian proof silver 75 rial coin and silver medal featuring the Cyrus Cylinder issued to commemorate the 2,500th anniversary of the founding of the Persian Empire.

Private collection.

appeared on banknotes of Reza Shah as early as 1938 featured on the two series of banknotes (the tenth and eleventh) that were issued in 1971 (1350) to commemorate the 2,500-year celebrations (Farahbakhsh 2005). Although the tomb of Cyrus featured again in the twelfth series (1975), the Cyrus Cylinder has never been shown on a Persian banknote.

Apart from Iran, stamps commemorating the 2,500-year anniversary were issued by Ethiopia, Romania, Nepal, Turkey, India, Pakistan, Umm al-Quwain, Ajman, Fujeira and Oman. The Cylinder itself appeared on the stamps of Ethiopia, Romania, Ajman and Oman (*figures* 41, 44–45), and the logo of the celebration incorporating the Cylinder, on a stamp of Umm al-Quwain (*figure* 46). Since the 2,500-year celebrations, the Cyrus Cylinder has appeared twice more on Iranian stamps. First, it was shown on a stamp commemorating the 22nd International Red Cross Conference held in Tehran in 1973. Beneath the Cylinder is written: 'Fut gravée sur ce Cylindre il y a 2500 ans la Proclamation de Cyrus le Grand. Genèse des Principes de la Croix-Rouge' (*figure* 47). Second, perhaps rather

44 Miniature sheet of stamps inscribed 'Ajman State' issued to commemorate the 2,500th anniversary of the founding of the Persian Empire.

Private collection.

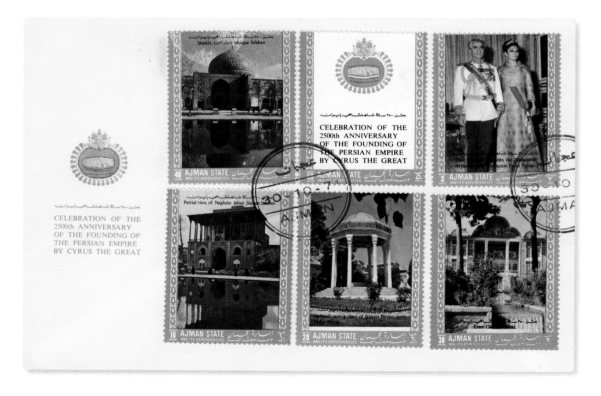

45 Miniature sheet of stamps inscribed 'State of Oman' issued to commemorate the 2,500th anniversary of the founding of the Persian Empire.

Private collection.

surprisingly, the Cylinder appeared on a stamp issued by the Islamic Republic of Iran in 2005. This was in the context of a miniature sheet of stamps (*figure 48*) issued to commemorate Expo 2005, which was held in Japan. The four stamps making up this sheet showed the Gateway of All Nations at Persepolis, wind towers at Yazd, typical Iranian desert architecture, and the Cyrus Cylinder, described on the stamp as 'The First World Charter of Human Rights'.

The suggestion that the Cyrus Cylinder might be loaned for a second time to Tehran was made by Neil MacGregor, the director of the British Museum, in April 2003 when he visited Tehran with John Curtis (then keeper of the Middle East Department at the British Museum) and Vesta Curtis (curator of Middle East coins at the British Mueum) to make preliminary enquiries about borrowing Iranian objects for the exhibition *Forgotten Empire: The World of Ancient Persia* that was held at the British Museum to great acclaim in the period 9 September 2005–8 January 2006. In due course, Iran made generous loans both to the *Forgotten Empire* exhibition and to the very successful exhibition *Shah Abbas: The*

CREATION OF AN ICON 91

46 (*far left*) Imperforate stamp inscribed 'Umm al Qiwain' issued to commemorate the 2,500th anniversary of the founding of the Persian Empire.

Private collection.

47 (*left*) Stamp featuring the Cyrus Cylinder issued by Iran on 8 November 1973 to commemorate the 22nd International Red Cross Conference held in Tehran.

Private collection.

Remaking of Iran (19 February–14 June 2009). The possible loan of the Cyrus Cylinder was mentioned on several other occasions during this period, but it was never an explicit condition of borrowing objects from Tehran. Nevertheless, the Trustees and director of the British Museum, mindful of the significance that the cylinder holds for all Iranians, were unanimous in their view that the Cylinder should be loaned to Iran, to give the Iranian public an opportunity to see it.

Consequently, John Curtis and Jill Maggs (loans manager at the British Museum) travelled to Tehran in the period 8–12 November 2009 to make arrangements for the loan. It was agreed in principle (subject to ratification by the Trustees) that the Cylinder would be loaned for the period 16 January–16 May 2010, and that the British Museum would also supply texts and illustrative material for a catalogue, labels and information panels. The opportunity was also taken for Jill Maggs to inspect the venue and check on security arrangements.

The British Museum was, however, obliged to reconsider its position when violent anti-government protests broke out in Tehran and other Iranian cities on *Ashura* (27 December 2009), the holy day on the tenth of Muharram (the Shiite mourning period). Resentment about perceived irregularities in the re-election of President Ahmadinejad had

48 Miniature sheet of stamps issued by Iran on 14 March 2005 to commemorate World Expo 2005 in Aichi, Japan.

Private collection.

been simmering since the previous June and now flared up with angry demonstrations throughout Iran.

By an extraordinary coincidence, around this time (in fact, on 23 December 2009 and 4 January 2010) two small clay fragments in the British Museum collection were identified by the late Professor W.G. Lambert and Dr I.L. Finkel respectively as belonging to a cuneiform tablet inscribed with part of the same text as the Cyrus Cylinder. The significance of these discoveries was immediately clear in that they showed the Cylinder was not a unique document and lent support to its identification as a proclamation. Faced with these two facts – what seemed to be escalating violence in Tehran and an important academic discovery that required urgent investigation – the Museum took the decision to postpone the loan until normality had been restored in Tehran and use the opportunity to organise an academic workshop on the new discoveries.

CREATION OF AN ICON   93

It has to be said this decision was not well received in Tehran, although Neil MacGregor was at pains to point out that the Trustees, in September 2009, had unanimously agreed to lend the Cylinder and had never departed from that position. There had, however, at the end of December 2009, been the violent demonstrations in Tehran, some of them explicitly directed against the United Kingdom, and in those circumstances the Trustees – whose supreme responsibility is for the safety of the collection – had reluctantly decided they must defer the loan. In spite of some frostiness, channels of communication were kept open and hopes were expressed on each side that the loan might still go ahead.

The planned workshop to discuss the new discoveries, organized by Irving Finkel, duly took place on 23–24 June 2010. Some of the papers delivered there form the basis of the present book. The workshop was judged by attendees to have been a very successful event, but, although there were some Iranian scholars present, there were no official representatives of the Iranian Cultural Heritage, Handicrafts and Tourism Organization (ICHHTO). Nevertheless, by late summer of 2010 the Museum felt the situation in Iran had stabilised sufficiently for the loan to be reactivated, and in the period 9–11 August 2010 Neil MacGregor went to Tehran with John and Vesta Curtis to resume negotiations. The group met with HE Hamid Baghaei, the vice president of Iran and Head of ICHHTO, and other Iranian officials, in the headquarters of ICHHTO in Khiaban-e Azadi on 10 August. After a certain amount of political posturing, cordial relations were re-established and the dates of 12 September 2010 to 12 January 2011 were agreed for the exhibition. The contract was then signed by Neil MacGregor on behalf of the British Museum and Mrs Azadeh Ardakani on behalf of the National Museum of Iran.

As the loan of the Cylinder to Tehran was such a momentous event, and as it attracted and continues to attract a good deal of public interest, we are taking the opportunity to describe the events surrounding the loan in greater detail than usual. The Cylinder and the two tablet fragments were transported by air in a metal case accompanied by John Curtis, Vesta Sarkhosh Curtis, who acted as translator throughout in negotiations and for speeches, Ken Uprichard (then head of conservation, British Museum) and Dean Baylis (senior administrator, Middle East Department, British

Museum). On arrival in Tehran at 4.31 a.m. local time on Friday 10 September 2010, the group was met airside by staff from ICHHTO and taken to the VIP lounge while passports were stamped and luggage collected. At 6.00 a.m. the party left Imam Khomeini International Airport in a cavalcade consisting of six vehicles with five motorcycle outriders. In Tehran itself the route was cordoned off and traffic had been stopped, so there were no hold-ups. The arrival at the National Museum at 6.40 a.m., where a crowd of about a hundred people had gathered, was filmed, and there were some interviews. The metal case with the Cylinder was then locked up in the strongroom. In the early hours of Saturday 11 September Neil MacGregor, Karen Armstrong (British Museum trustee), Joanna Mackle (director of public engagement, British Museum) and Birgit Brandt (representing the British Academy) arrived in Tehran for the opening ceremony that was scheduled for the next day. In the meantime, however, there was an 'inspection' or 'authentication' ceremony on the morning of Saturday 11 September, the purpose of which was to allow a panel of experts to examine the Cylinder, sign a condition report and certify that it was genuine. On the Iranian side there were nine experts, who all declared themselves completely satisfied that the Cylinder was indeed the original object and not a replica, and the ceremony was witnessed by Mr Baghaei and other senior members of the ICHHTO. After all the experts had

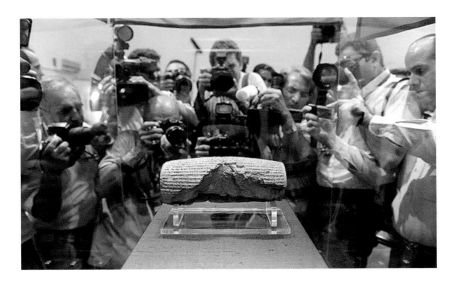

49 Photographers massed behind the case with the Cyrus Cylinder, National Museum of Iran, 11 September 2010.

50 General view of the exhibition hall with the Cyrus Cylinder in the National Museum of Iran, Sepember 2010.

Photo J.E. Curtis.

been given a chance to inspect the Cylinder (but not handle it), it and the tablet fragments were placed in the exhibition case, which was then locked. This ceremony took place in the presence of a large number of journalists, photographers and camera crews (*figure* 49).

The Cylinder and the two tablet fragments were exhibited in a freestanding square case with fibre-optic lighting, which was sent from the British Museum. This case was stationed in the middle of the gallery formerly known as 'the treasure room', a large room on the upper floor at the front of the museum (*figure* 50). A low glass wall around the case at a distance of about 50 cm from it prevented crowds from getting too close. Around the walls of the room were eleven explanatory panels in Persian and English describing the Cylinder, its discovery and its significance; the texts, with illustrations, had been sent from London. Outside the exhibition hall, the entire upper floor of the museum had been given over to a re-creation of Persepolis in photographs and casts, and the staircase to the upper floor was lined with a photographic frieze showing figures, priests or servants, mounting stairs at Persepolis. The whole exhibition was designed with taste and flair, and was a remarkable demonstration of how an entire exhibition can be made from a single object, if that object is powerful enough. The visitor experience even extended to the outside of the museum. In the forecourt were banners and an ice-cream parlour and two souvenir stalls selling mugs, replicas, posters and other memorabilia (*figure 51*).

The opening ceremony in the afternoon of Sunday 12 September 2010 started with a reception in the Islamic Museum at which about 30 ambassadors were present, including Simon Gass from the UK. Neil MacGregor and John and Vesta Curtis were then invited to join the presidential party for the opening of the exhibition, and on the steps of the National Museum were introduced to the president of Iran, HE Mahmud Ahmadinejad, HE Esfandiyar Rahim Mashaei, adviser to the president, and Mr Baghaei. The group then proceeded upstairs to the exhibition and stood in a line in front of the case, with photographers massed on the other side of the case (*figure 52*). The case was covered with a small 'tent' made from the Iranian flag, and as everybody watched the tent was drawn upwards by a cord, revealing to great applause the Cyrus Cylinder inside the case. The group then proceeded to the auditorium in the Islamic Museum for an opening ceremony that lasted from 3.30 until 6.00 p.m. The stage was decorated in Persepolitan style, with a reconstructed doorway from the Hall of 100 Columns on the left and on the front of the stage was a file of guards with rosettes and crenellations. A huge replica of the Cyrus Cylinder was on the right of the stage.

51 The forecourt of the National Museum of Iran with installations for the Cyrus Cylinder exhibition, September 2010.

Photo J.E. Curtis.

CREATION OF AN ICON 97

52 British Museum staff with President Ahmadinejad at the opening of the exhibition in Tehran, 12 September 2010.

The ceremony began with a musical performance with a *daf* (a type of Persian frame drum), which was followed by five speeches. Mrs Azadeh Ardakani, the director of the National Museum, began by introducing the exhibition, thanking the British Museum for the loan of the Cylinder, and remarking how honoured she was that the Cylinder would be on exhibition in the National Museum. She was followed by Neil MacGregor, the director of the British Museum, who said that the British Museum was very grateful for generous loans from Iran to the *Forgotten Empire* and *Shah Abbas* exhibitions, and in return the Trustees were glad to have the opportunity of lending the Cyrus Cylinder to Tehran. The Cylinder described how Cyrus had restored shrines and repatriated deported peoples. Because of these enlightened acts, which were rare in antiquity, the cylinder had acquired a special resonance and was valued by people all around the world as a symbol of tolerance and respect for different peoples and different faiths. The Trustees recognised that the Cylinder was part of a shared heritage belonging to all humanity, and for this reason they felt they had an obligation to allow it to be shown to as many people around the world as possible, including Iran, the birthplace of Cyrus the Great.

Mr Baghaei recalled the destructive war waged by the Assyrian king Ashurbanipal (668–631 BC) against Susa and contrasted his treatment of captives and intolerant attitude with the enlightened approach of Cyrus the Great when he captured Babylon. He also compared Cyrus with Abraham Lincoln: they had both rejected slavery, but Lincoln more than 2,000 years later, and he had been murdered for his efforts. In a speech replete with mystical overtones, Mr Mashaei extolled the virtues of Cyrus and emphasised the importance of his proclamation (i.e. the Cylinder) for the modern world with its values of peace, goodness and justice. President Ahmadinejad described Cyrus as a protector of monotheism who allowed the Jews to return to their homeland. In his Cylinder he proclaimed freedom and liberty, values that had been important throughout Iranian history, while at the same time protecting Iran from foreign aggressors.

This theme of protecting Iran was taken up in a performance featuring a succession of actors playing the parts of Cyrus the Great (wearing a costume dating from the time of the 1971 celebrations); of Kaveh Ahangar, the blacksmith of the *Shahnama* (the Iranian national epic by Firdowsi) who rescued Iran from the usurper Zahak; and a martyr (*basiji*) with the flag of the Islamic Republic of Iran, who repelled invaders in the Iraq–Iran war. The president presented to each of the actors in turn a black-and-white *keffiyeh*, the traditional symbol of Palestinian resistance. During the performance representatives of various Iranian tribes such as Qashqais and Lurs also appeared on the stage. After the dramatic presentation there was a film, which was subsequently shown outside the gallery throughout the exhibition. This showed an Achaemenid soldier riding through Persian history in search of the Cyrus Cylinder, which he eventually found on exhibition in the National Museum. At the end of the ceremony, the audience was entertained by an Iranian singer. Contrary to expectations in view of his heavy workload, President Ahmadinejad stayed throughout the ceremony.

In the evening a dinner for about 70 people was hosted by Mr Baghaei in the Golzar Hall of the Laleh International Hotel. Apart from the British Museum party and the British ambassador and his deputy Jane Marriott, there were a number of foreign ambassadors and senior officials from ICHHTO and elsewhere. There were brief speeches from Mr Baghaei and Neil MacGregor, both translated by Vesta Sarkhosh Curtis.

Throughout the duration of the exhibition there were a number of visits to Tehran by British Museum staff (John Curtis four times, Vesta

Sarkhosh Curtis three times, Ladan Akbarnia once) to check on the state of the installation, but this was strictly unnecessary as the Cylinder was scrupulously looked after and the exhibition beautifully maintained by the Iranian colleagues. The great success of the exhibition did, however, lead the Iranians to ask for an extension so that the Cylinder could be kept in Tehran for the Now-Ruz holidays. This was agreed to, subject to the signing of a memorandum of understanding as follows:

1. The British Museum agrees to extend the Cyrus Cylinder exhibition until 15 April, subject to the signing of a new loan agreement.
2. There will be no further extensions after 17 April.
3. Both sides agree to use their best endeavours to support in their own countries the cultural, archaeological and museological projects of the other party.
4. Both sides agree to investigate, encourage and support futher loan programmes between Iran and the UK.
5. The National Museum of Iran agrees to support the Parthian Coin Project.
6. The ICHHTO recognises the important work done by the British Academy and its sponsored institutions to promote Iranian Studies.

At the end of the exhibition, the group that went from the British Museum to attend the closing ceremony and escort the Cylinder back to London consisted of John and Vesta Curtis, Ken Uprichard and Dean Baylis (couriers), and Neil MacGregor and Joanna Mackle with John Wilson (BBC radio, *Front Row*) and Ben Hoyle (*The Times* newspaper). The last public day of the exhibition was Friday 15 April, and on the following day was the closing ceremony (Saturday 16 April), again in the auditorium of the Islamic Museum. The stage was now flanked by replica Persepolitan winged sphinxes and there was a frieze of lotus flowers along the front. There were speeches from Mrs Ardakani and Neil MacGregor, followed by a musical interlude. The music was played on Elamite-type musical instruments reconstructed on the basis of cylinder seals, by five musicians (four men and one woman), several of whom sang. There were songs about the Cyrus Cylinder (*Manshur-i Kurosh*) and other pieces. After the concert there was a speech from Mr Baghaei, who started his talk by saying the message of the Cylinder had gone unheeded in many parts of the world. He quoted Goethe as saying it was remarkable that a country that had been invaded and oppressed so many times had still managed to maintain its national identity. He ended his speech with a quotation from the last part of Firdowsi's *Shahnama*: 'If there is no Iran then my body won't exist...' Mr Baghaei was followed by a recital in the ancient tradition of Persian storytelling by Amir Sadeghi, who read in a powerful and booming voice a poem that he had written himself in the style of Firdowsi's *Shahnama* about the Cyrus Cylinder. Mr Sadeghi was dressed in the style of Firdowsi in a turban and white coat, carrying a staff. Lastly, there were some presentations to Neil MacGregor and to Iran Museum staff who had been involved in the exhibition, and a new high-quality cast of the Cyrus Cylinder recently made by the British Museum was presented to Mrs Ardakani for the National Museum.

The audience then proceeded to the steps of the Museum, for the unveiling of a brass plaque that had been set into a tiled surround at the back of an alcove on the left-hand side of the entrance hall (*figure* 53). The newly presented replica of the cylinder was placed on a bed of sand at the bottom of the alcove, which was sealed off with a glass panel. The brass plaque is inscribed in Persian and English. The gist of the English version is as follows:

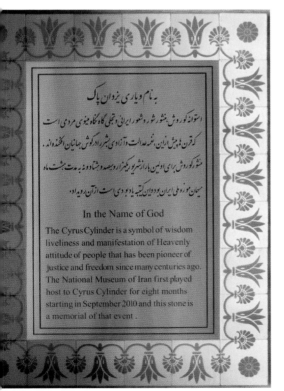

53 Brass plaque commemorating the exhibition and a cast of the Cyrus Cylinder in the porch of the National Museum of Iran, April 2011.

Photo J.E. Curtis.

In the name of God. The Cyrus Cylinder is a symbol of wisdom and good living and a manifestation of the heavenly attitude of a people that pioneered justice and freedom many centuries ago. The National Museum of Iran hosted the Cyrus Cylinder for eight months starting in September 2010 and this plaque commemorates that event.

Guests then went up to the exhibition hall where many photographers were waiting; the Cylinder was removed from its showcase with great ceremony and placed in the metal carrying case, which was then taken to the strongroom and locked up. At this point Mr Mashaei arrived late, apologising profusely for having missed the ceremony. He asked if the Cylinder could be brought back to the gallery, which was agreed. In its presence Mr Mashaei delivered a long speech, which was reported in next morning's newspapers. He again spoke about the values embodied in the Cylinder. Although

CREATION OF AN ICON   101

regretting that it was about to leave Iran, he supported the principle that it should be exhibited in as many countries as possible to give more people the chance to see it. The Cylinder was then returned to the strongroom, after which Mr Mashaei held court with journalists and reporters in the large room, previously a lecture hall, on the upper floor of the Museum. In the evening the British Museum party was treated to dinner at a restaurant close to the former Shah's palace in Saadabad in north Tehran. Neil MacGregor and Joanna Mackle left Tehran in the early hours of the next morning, while the four couriers stayed on for another twenty-four hours so that a condition check of the Cylinder could be made by Ken Uprichard and Miss Mahnaz Gorji, head of conservation in the National Museum of Iran. The drive back to the airport in the early hours of Monday 18 April was conducted with the same stringent security arrangements as the arrival in Tehran, and, although the flight was delayed by nearly five hours, because it was Army Day and civilian flights were temporarily grounded, no difficulties were encountered. In fact, the passage through immigration and customs was swift and trouble-free. There was no attempt to block the export of the Cylinder, nor any suggestion that it should not be returned to Britain. The only suggestion that the Cylinder should be impounded had come in a newspaper report at the beginning of the exhibition, but it had not been repeated. The Iranian officials themselves behaved impeccably throughout.

The exhibition in Tehran was extremely popular, but estimates of the total number of visitors have varied considerably. The official figure is just over 1 million but other estimates are rather lower. According to the *Tehran Times* of 10 January 2011 (quoted in Wikipedia) 190,000 people had visited by 10 January 2011, with 48,000 of them coming within the first ten days (*Tehran Times*, 26 September 2010). If that figure were projected for seven months, it would give a rough total of 330,000. The facts are that generally people were allowed into the exhibition in groups of 15, with each visit lasting five minutes. If the numbers stayed the same throughout the exhibition and there was full capacity this would give a total of approximately 465,000 visitors. However, we were informed that on one day alone in September (children's day on 10 Mehr) there were more that 10,000 visitors, so clearly groups were not always restricted to 15 people each time. In the light of this, it seems reasonable to conclude that the exhibition was seen by up to half a million people. Two publications were available at the exhibition: the official exhibition catalogue, in Persian and English, containing the panel texts that had been sent from London (Ardakani and MacGregor 2010), and a book by Dr Shahrokh Razmjou, also in Persian and English, with a rather fuller account (Razmjou 2010).

At the time of writing, plans are at an advanced stage to send the Cyrus Cylinder accompanied by a small selection of iconic Achaemenid objects to five different venues in the USA in the course of 2013. This is in line with the policy of the British Museum Trustees to make important parts of the collection accessible to audiences worldwide, recognising that there is keen interest in the USA in the Cyrus Cylinder among expatriate Iranian, Jewish and other groups.

NOTES

I am grateful to my wife Vesta Sarkhosh Curtis for reading a draft of this article and making helpful suggestions, and for translating sources in Persian.
1. I am grateful to Dr Shahrokh Razmjou for this information.
2. Personal communication from Mr Abolala Soudavar.
3. See his article 'Cyrus, King of Persia', *Acta Iranica* I (1974), pp. 127–34.
4. The title of the conference was 'Continuité de la culture iranienne'.

5

The Cyrus Cylinder: a Persian perspective

SHAHROKH RAZMJOU

Babylon before the arrival of Cyrus

At the point in 539 BC when the Persian army of Cyrus the Great marched through the city gates of Babylon the king on the throne was Nabonidus (Babylonian Nabu-na'id), whose 17-year reign had begun in 556 BC. Nabonidus was in many respects an unconventional Babylonian ruler, and by the end of his reign Babylonia was in a state of social and religious disarray, which Cyrus was quick to exploit.

How Nabonidus came to the throne of Babylon is still obscure. He became king following a string of troubled or ineffective rulers whose achievements were negligible in comparison with those of the founder of the dynasty, Nabopolassar (658–605 BC), and more especially his son and successor, the great Nebuchadnezzar II (605–562 BC). It was Nebuchadnezzar who established Babylon as the greatest power in its more than thousand-year history, both politically and architecturally, but the great empire that he bequeathed was thereafter in lesser hands, and Nabonidus, once acknowledged as king, was largely preoccupied with matters other than empire-building.

Nabonidus, then a court official already in his sixties, usurped the throne by leading a conspiracy in which Labashi-Marduk, the child king of Babylon, was deposed and murdered after a few months' reign. How complex or bloody the process might have been, we do not yet know.

Virtually nothing is known about his father beyond his name, Nabu-balatsu-iqbi, who might well have been an Assyrian official. Nabonidus made no attempt to associate himself with the previous dynasty, but established his own, known as the dynasty of Harran after the city of his mother and the location of the temple of Sin, the moon god (*Dynastic Prophecy* ii: 12; Grayson 1975b: 33). The dominant figure in his background was certainly his mother, Adda-guppi, who lived to the venerable age of 104, and has become well known thanks to her autobiographical inscriptions, carved posthumously on twin stelae installed in the temple of Harran in modern Turkey. Adda-guppi in Harran shows herself to have been an ardent devotee of the moon god Sin. We know that Harran was sacked by the Babylonians and the Sin temple left in ruins, but her account omits this military event, saying that Sin, king of gods, became angry with his city and temple and went up to heaven. Adda-guppi describes how, in a dream, the moon god Sin chose her son Nabonidus to rebuild the temple of Harran and, we can be sure, installed in the young man what was to be a lifelong devotion to Sin and the desire to promote his cult. Nabonidus himself declared that the previous dynasty had become weak and that he was chosen by divine will to rebuild the moon god's temple at Harran. Adda-guppi's personal devotion to the moon god was thus to be instrumental in changing the destiny of Babylon for good.

Marduk was the patron god of Babylon and the great god of the Babylonian world, promoted explicitly and vociferously by Nebuchadnezzar, with long-standing worship and a flourishing cult in the temples. Theologians at this time were even arguing that Marduk was more than king of the gods – as he had been for centuries – but rather the only god, recasting the other leading gods of the pantheon as mere aspects of Marduk. Sin, in contrast, was a less familiar deity to the Babylonians and his cult was not widely practised in Babylon. Nabonidus proved to be less than sympathetic to the Babylonian Marduk cult, perhaps partly fuelled by the earlier destruction of Harran and the Sin temple by the Babylonians. Rather he devoted himself totally to the cult of Sin at the cost of Marduk and his cult. The inevitable consequence was reduced concern on the part of the king for the welfare of Babylon itself, which was irradicably 'in the hands' of Marduk, and his desire to impose Sin worship in general would be resisted at every

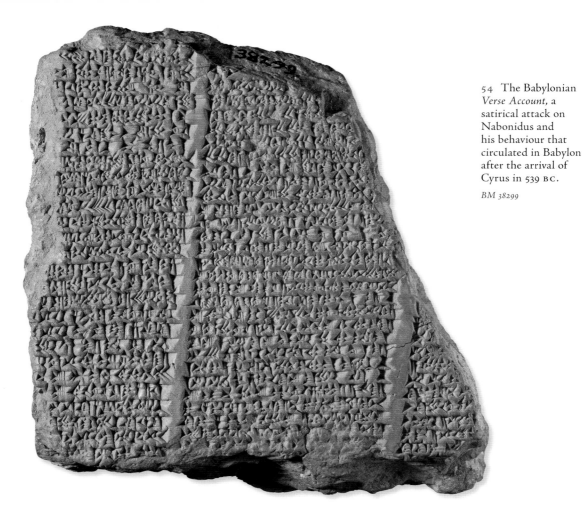

54 The Babylonian *Verse Account*, a satirical attack on Nabonidus and his behaviour that circulated in Babylon after the arrival of Cyrus in 539 BC.
BM 38299

turn. Controversially, he installed a statue of the moon god within the central Esagil temple, the home and sacred precinct of Marduk. This was to confront the Babylonian religious sensitivity directly. A later satirical pamphlet, the Babylonian *Verse Account*, described the statue of Sin as

> crowned with a tiara ... its appearance is [that] of the eclipsed moon ... When he [Nabonidus] worshipped it, its appearance became like that of a demon crowned with a Tiara ... [and] his face turned hostile. (*Verse Account* col. i 28–31: Oppenheim 1969: 313; Schaudig 2001: 567)

Even worse, he ordered the construction of a major temple for Sin in Babylon that imitated the temple of Marduk.

There were probably mixed reasons that led Nabonidus to the major step of abandoning his capital, but this religious struggle was certainly a crucial factor. In, probably, 553 BC, Nabonidus appointed his first-born son, Belshazzar (Bel-šar-uṣur), as regent and entrusted to him the ruling of Babylon. He himself, with retinue, priesthood and a substantial military force, left the city for Teima, an important oasis in far distant Arabia. The Babylonian *Verse Account*, quoted above, tells us how he conquered that city, massacring inhabitants and imposing hard labour upon the survivors, including women and young people, who had to build a palace for the incoming king and a temple for his god Sin. Crucially, Nabonidus settled at Teima, not returning to Babylon for ten years. The inhabitants of Babylon must have felt themselves completely neglected for the whole of that decade, although Belshazzar as regent seems to have maintained stability and not followed his father's rejection of Marduk. In the absence of Nabonidus, however, the New Year Festival, which depended on the presence of the king to 'take the hand of Marduk' and thereby establish well-being in the country for the coming year, had repeatedly to be cancelled. The *Nabonidus Chronicle* (Grayson 1975a: 106) recorded bitterly:

> Nabonidus, the king, stayed in Teima.… The king did not come to Babylon for the ceremony … the god Nabû did not come to Babylon, the god Bel did not go out of Esagila in procession; the festival of the New Year was omitted.

Theologically a single instance like this would have been an ominous development, and its repetition could only result in widespread and growing discontent among the clergy, who were chronically fearful as to its effects. Marduk and other gods would be angry, and no doubt many other essential rituals of the Babylonian year would be abandoned, or expected to lose their efficacy.

It is not clear why Nabonidus decided to return to Babylon after his ten years' absence. It might have been in response to escalating tensions in the city, but it is more likely that reports of the rising power of Cyrus close to his borders after the fall of Lydia in 546 BC were proving too serious to ignore. The king's return to his capital, however, did not improve the situation, and to many he did not even seem to be mentally stable. The Babylonian *Verse Account* describes how Nabonidus stood up in the Assembly and praised himself as wise and able to see what is hidden despite being unable write with a stylus; one very obscure god – named Ilte'ri – had shown him secret things in a vision so that he was aware of wisdom and everything else (*Verse Account* col. iv; Oppenheim 1969: 314). He was even described by the Babylonians as 'mad'.

This condemnation of King Nabonidus as a mad person, which by Babylonian standards was justifiable, was transferred in later tradition to his famous predecessor Nebuchadnezzar, largely through the writings of the biblical prophets. The latter's crime of destroying the Temple in Jerusalem and driving all the Judaeans into exile in his capital condemned him forever in biblical terms, culminating in the description of his madness in the Book of Daniel, and ultimately the iconic image of the king, on hands and knees, painted by William Blake (Finkel and Seymour 2008: 166–9).

According to the remarkable Babylonian text known as the *Dynastic Prophecy* the life of Nabonidus was spared after the Persian conquest, and he was packed off to another land (Grayson 1975b: 33, ll. 20–21; van der Spek 2003: 316). There is apparent confirmation of this from Berossus, the Babylonian writer and priest of the Hellenistic period, according to whom Cyrus treated Nabonidus kindly and sent him to rule as a vassal king in Carmania, modern-day Kerman (Berossus, quoted by Josephus, *Contra Appionem* I: 20–21). (This is reminiscent of the Greek accounts of making Astyages a regional governor, as discussed below.) The *Nabonidus Chronicle* makes reference to the 'death of the king's wife' about four months afterwards, and this may refer to the wife of Nabonidus. Apparently there was an official mourning period for her in Akkad under Cyrus.

Cyrus in Persia

By the time Cyrus led his armies through the gates of Babylon in 539 BC he was well on his way to establishing the great empire that resonated through history. Little enough is known of Cyrus' life before he became king. The Greek authors wrote extensively about the subject, but most of their accounts are mixed with myth and legend, and differing stories and legends were soon to circulate among the peoples of the ancient world. The most extensive and detailed accounts are by Herodotus and Xenophon.

According to Herodotus, Cyrus was the son of Mandana, daughter of the Median king Astyages, and of Cambyses, king of Persia. He was thus the grandson of the last king of Media. When Mandana was pregnant with Cyrus, Astyages dreamt that a vine grew out of his daughter which covered the whole of Asia. The magi and dream-interpreters in his court warned him that his daughter would give birth to a child who would overthrow him and rule in his place. After Mandana gave birth, Astyages ordered Harpagus, his chief adviser, to take the baby away and kill him. Harpagus instead gave him to his cowherd, whose wife had just lost her own baby. They kept the boy and displayed their own dead baby to a trusted bodyguard as a proof of mission accomplished (Herodotus *Histories* i.108–13). The boy grew up in the cowherd's house until the secret was revealed, whereupon Harpagus was dreadfully punished by Astyages for disobeying his orders.[1] Young Cyrus was sent to his father in Persia until he rebelled against the rule of Astyages. The Median king decided to solve the problem militarily by sending his army against Persia, but the Median army – headed by Harpagus – joined Cyrus. The united Persian and Median armies marched together to Ecbatana and arrested Astyages (Herodotus *Histories* i.127–8).

This dramatic story of Herodotus became the most popular version of the life of Cyrus for obvious reasons. However, it merely exemplifies the genre of a historical figure, abandoned in childhood and raised in a different environment before achieving great deeds, and is far from real history.[2] The story told by Xenophon is different but also refers to how Cyrus was trained in justice and how he became popular among the Medes and the Persians.

Babylonian texts confirm that Median forces joined up with Cyrus. The *Nabonidus Chronicle* reports that in 550/549 BC, King Astyages (here called Ištumegu) mustered his troops and marched against Cyrus, king of Anshan, but his own army revolted against him and they delivered him to Cyrus (*Nabonidus Chronicle* col. ii: 1–2). Astyages' offensive seems to have been in response to the growing threat of Cyrus and his kingdom. Afterwards Cyrus marched to Agmatanu (Ecbatana, modern Hamedan), the Median capital. The classical sources claim that Astyages' life was spared. According to Herodotus, Astyages was maintained in his own court for the remainder of his life; other Greek sources record that Cyrus made him governor of a region. Although various explanations are possible for the revolt of the Median army against Astyages, Cyrus' charisma and popularity among the Medes must have been crucial in persuading them that he, a Persian, was preferable to their own king, leading to the capture of their own capital. We do not know enough to explain such a complex sequence of events, but it was undoubtedly the beginning of Cyrus' meteoric rise and the formation of his empire.

Cyrus, king of Anshan

Cyrus describes himself in the Cylinder inscription (lines 20–22) as king of Anshan, and the same prestigious title is applied to his ancestors:

> I am Cyrus, king of the universe, the great king, the powerful king, king of Babylon, king of Sumer and Akkad, king of the four quarters of the world, son of Cambyses, the great king, king of the city of Anshan, grandson of Cyrus, the great king, king of the city of Anshan, descendant of Teispes, the great king, king of the city of Anshan.

The names of Anshan and Susa had been familiar to the people of Mesopotamia since ancient times as the two major Elamite centres. Susa was located in the low-lying southwestern plain of Iran, and Anshan in the eastern highlands. A trade route to the east had stretched through the territories of Anshan since very remote times.[3]

Persians speaking an Indo-European language migrated to and settled in the neighbourhood of Anshan, which came to be named after them (i.e. Parsua).[4] They seem to have enjoyed peaceful relations and coexistence with the Elamites. Soon, relations were sufficiently close to unite them as allies against the invading Assyrians. At some point in the seventh century BC, probably under the Persian king Teispes (reigned c. 675–c. 640 BC), the Elamite rulers of Anshan were replaced by a Persian royal dynasty. It is still unclear how and when Anshan became the capital city of the Persians, but afterwards rulers of this kingdom held the traditional title King of Anshan. However, in the *Nabonidus Chronicle* (ii.15) Cyrus is also called by a different title, king of Parsua (Persia).

According to the evidence, Cyrus was in fact the second king in this dynasty with the name Kuraš or Kuruš. A royal seal of his grandfather, Cyrus I (c. 640–c. 600 BC), was reused more than a century later on some clay documents at Persepolis, showing a victory scene of Cyrus I on horseback against an unidentified enemy, with an accompanying inscription in cuneiform introducing him as 'Cyrus of Anshan, son of Teispes'.

For a short period after the conquest of Babylon, Cyrus still used the title 'king of Anshan', as we see in inscriptions from Babylon and Ur (Schaudig 2001: 549), but thereafter that title was no longer used by him. Instead Babylonian titles such as 'king of the lands' (in Uruk; see Schaudig 2001: 548) were preferred, clearly reflecting Cyrus' greater status and self-awareness. This title is often applied to him in non-royal inscriptions too. The reason is self-evident; any king who ruled over Persia, Media, Lydia and many other regions from Central Asia to the Black Sea and the Mediterranean to the gates of Africa, with Babylon as the metropolis of the ancient world, was in fact king of most of the then civilised world. Therefore, as the ruler of many lands, peoples and civilisations, he was fully entitled to adopt the traditional titles of the rulers of the Ancient Near East, such as 'king of kings', 'king of the universe' or 'king of the four quarters'. The conquest of Babylon completed the transformation of the Persian kingdom into a world empire and Cyrus emerged as the Great King ruling over vast dominions. Therefore Cyrus was now 'king of kings' and there is no further reference to the title 'king of Anshan' for Cyrus in Mesopotamian documents.[5] There was no longer a king of Anshan, which was only one city among many others in the empire. The title 'king of Anshan' became part of history. Incomplete and limited archaeological excavations on a small scale at Anshan (modern Tal-e Malyan in Fars province, Iran) have not yet provided evidence of any Persian royal monuments of the ruling dynasty.

Cyrus and Pasargadae

With the capture of Babylon and the rapid formation of the Persian Empire Anshan was no longer suitable as Cyrus' capital; nor could Babylon itself, despite its status, function as the Persian centre. The latter was, however, always to be an important city; Cyrus' son Cambyses as prince regent was appointed governor/king of Babylon to learn kingship in accordance with old tradition.[6]

Cyrus founded his own capital city of Pasargadae (called *Batrakatash* in Elamite texts from Persepolis), far from Babylon, in the heart of Persia. Strabo (xv.3.8), as well as Justin and Nicholas of Damascus, believed that Cyrus built this city at the spot where he had been victorious over Astyages, although more practical factors tend to operate when choosing a location to found a city.

This royal city consisted of palaces, built in white limestone and mud brick, with columned halls whose columns were made of a combination of white and black stones. There was also a stone tower with an unknown function (modern Zendan-i Suleiman). This complex was built on the plain next to the river Polvar. The whole site could be seen from a fortified citadel (modern Tall-e Takht), built on a huge stone platform on the top of an overlooking hill. The shape and plan of the buildings and palaces at Pasargadae were designed in a new style, totally different from the former palaces in the region and introducing a new form of royal architecture to the Ancient Near East. They mostly followed the older Iranian tradition of columned halls, and some also included the artistic styles of other cultures. At Pasargadae, the columned halls were constructed with column capitals in the shape of double lions. The columnar architectural style was later developed and reached its peak in the architecture of Persepolis and Susa.

The palaces and gatehouse at Pasargadae were surrounded by royal gardens, called 'paradise'. This Persian word for walled gardens was also used later for the heavenly garden and was taken into other cultures. These were rectangular gardens, divided by a sophisticated drainage system into four quarters, reminiscent of later Persian *chahar-baghs*, or four-gardens (Stronach 1990). The gardens were watered by channels, which had a number of small pools in between (Stronach 1978: 105–12). Creating such paradises was an Iranian tradition, strongly encouraged by the kings, princes and governors all over the empire.[7]

Porticos of the palaces faced these amazing gardens. Members of the royal family, dignitaries and visiting guests could enjoy the view and fragrance by sitting on benches made of white and black limestone covered with fine textiles, and running all around the porticos. This must have impressed visitors and foreign guests and given them an unforgettable experience of peace and harmony in a heavenly garden on earth. The

55 Photograph of the tomb of Cyrus at Pasargadae, taken in about 1965.

Photo: P. Gotch.

tranquility of Pasargadae may reflect a part of the ideology and mentality of Cyrus as a ruler. This may also recall the personality of another Persian prince, Cyrus the Younger and his concern for gardens (d. 401 BC), who himself designed and worked on his own paradise in Sardis and told Spartan general Lysander that he never dined before working in the garden (Xenophon *Oeconomicus* 4.20–22).

The mausoleum of Cyrus, the most significant structure in the royal city, was also, as reported by Greek historians, originally surrounded by gardens (*figure 55*).[8]

There is little textual evidence from the site referring to the name Cyrus. However, short trilingual inscriptions carved on the buildings and the remaining reliefs briefly introduce Cyrus as an Achaemenid king (CMa–c; Kent 1953: 107, 116).[9] Although some scholars believe that the inscriptions were made later by another Achaemenid king, in either case they connect the complex and the site with Cyrus. Arrian, quoting Aristobolus, refers to an inscription on the tomb of Cyrus (Arrian *Anabasis* 6.29; also Strabo *Geography*: 15.3.7; Plutarch *Alex.* 69.4) in which Cyrus speaks in a humble way to the future visitors to his tomb,[10] but no evidence of such an inscription survives.

With the accession of Darius the Great (522–486 BC) to the throne, Pasargadae lost its status as a capital city. New capital cities of Susa and Persepolis were constructed and replaced Pasargadae; the city, however, retained its reputation and importance for the rest of the Achaemenid period. The Greek sources refer to the coronations of Persian kings at Pasargadae after Cyrus, at which ceremonies his robe was worn as a part of the ritual (Plutarch *Artaxerxes* 2.3). This shows its role as a highly important memorial city linked with Cyrus. Until the end of the Achaemenid period the tomb of Cyrus was well protected by the magi, but it was looted at the time of the invasion of Alexander.

The conquest of Lydia

In Lydia, in Asia Minor, Cyrus' overthrow of his rivals, the mighty Medes, was probably good news for the Lydian monarch, and seen as a great opportunity to claim Median territories. But Cyrus as the successor of the Medes was not an easy target. The campaign of Croesus, king of Lydia, against Cyrus was disastrous. Croesus was defeated and Cyrus followed him to his capital Sardis, close to the shores of the Aegean Sea. Sardis was taken and Croesus captured.[11] Some sources (e.g. Herodotus) say that Cyrus made the defeated king his consultant. The fall of Lydia brought the borders of the Persian Empire to the shores of the Mediterranean and the Black Sea. It also alerted Babylon and Nabonidus, king of Babylon (556–539 BC), to the existence of a new power.

Babylon and Cyrus, sources and views

Accounts of the fall of Babylon to the Persians and the installation of its new ruler must have rapidly crossed many borders and been heard by numerous nations. The news would have been carried by travellers and merchants, and also sent in the form of official documents to other lands to be read aloud in public.[12]

In Egypt, an ally of Babylon, the news was received with anxiety and met with silence. After the peripheral regions of Babylon acknowledged Cyrus as their legitimate ruler, the pharaoh Amasis II (570–526 BC) experienced concerns about the rising power beyond his eastern borders. The fall of Babylon brought this new neighbour right up to the Asian boundaries of Egypt.

For the Greeks it was a different matter. Babylon was to many little more than an exotic name, a remote entity in the east that had fallen into the hands of another eastern monarch, although it might have been surprising to some that the mighty Babylon had fallen so easily. Later, when Greek historians sought to tell the story of the fall of Babylon, they mixed it with other stories for their Greek audience. Herodotus recounts that Cyrus ordered a canal to be dug and the River Euphrates diverted (*Histories*: 1.189–91). While the Babylonians celebrated a festival and were busy dancing, the army of Cyrus marched quietly into Babylon along the river bed. The theme of Herodotus' story seems to be influenced by stories such as the sack of Troy, and indeed the fall of Babylon came to be seen in the ancient world as equally important. Herodotus records that some of the inhabitants did not even notice that Babylon had been captured (Herodotus *Histories*: 1.190–91), which is hard to believe, but may ultimately reflect the peaceful way in which the city had been captured. Xenophon's story is influenced by Herodotus but adds military clashes during the capture of the city (Xenophon *Cyropaedia*: 7.5.1–36) to make the story more exciting and entertaining.

Obviously the Greek versions should scarcely be valued as history in the modern sense, intended as they were for the entertainment of an audience who preferred action and drama to a mundane story of peaceful occupation. The Greek authors neither witnessed the events themselves nor were able to interview any direct witnesses, as they lived centuries later. For these reasons their accounts stand in contrast with the first-hand textual evidence, found in Babylon.

No written evidence survives to recount the Persian history of this event. According to Herodotus, 'the Persians were well informed in history' (Herodotus *Histories*: 1.1) and certainly knew their own. No evidence of their records, such as 'chronicles' (mentioned in Ezra: 6:1–2; Book of Esther: 2:23, 6:1, 10:2) and *basilikai diphtherai*, 'royal records' (Diodorus 2.32.4, referring to Ctesias; Briant 2002: 6), has yet been recovered. This could well be due to Macedonian looting of the archives and burning of records which were written on perishable material,[13] as well as to the accident of archaeological discovery. Our understanding of the historical event is wholly dependent on the Babylonian records.

Babylon had a long tradition of recording political, military and other events. Some of the most useful of these sources were largely independent of political influence, such as astronomical diaries and (although to a lesser extent) chronicles, which had their own format for recording data as accurately as possible with minimum error. The events were recorded simultaneously in different texts by Babylonian scribes. Apart from the Cyrus Cylinder itself – which is more of a royal decree and contains the words of Cyrus himself – detailed and important information about the fall of Babylon is found in several other texts. These include the *Nabonidus Chronicle*, the *Verse Account*, which is more politically oriented, and the text known as the *Dynastic Prophecy*.

The tablet containing the *Dynastic Prophecy*, a Seleucid period copy of which is now in the British Museum, refers to Cyrus as an Elamite king (see above) who will arise and remove the rebel prince (Nabonidus) who plotted evil in Akkad, and who 'will cause the land of Akkad to sit in a dwelling of peace'.[14]

The Fall of Babylon

In 539 BC the united Persian and Median armies of Cyrus marched on Babylon. The detailed description of the fall of the city, as described in the above-mentioned texts, allows a full reconstruction of the historical event.

Babylonian texts report that Nabonidus took his army and went to meet his advancing opponents. The two armies clashed at Opis, north of Babylon. The army of Nabonidus was defeated in this battle and was forced to retreat.[15] On the fourteenth day of Tashritu (10 October), 539 BC, the city of Sippar was taken by Cyrus without a fight (*Nabonidus Chronicle* iii: 12–14). Nabonidus fled, but had to make his way back to Babylon. Apparently, Cyrus had sent Gubaru,[16] his general and governor of Media, with part of his army to proceed at once to Babylon, before Nabonidus could reach the city. Two days later on the sixteenth of Tashritu (12 October) the army of Cyrus, under Gubaru, entered Babylon 'without a battle' (*Nabonidus Chronicle* iii: 15). No resistance is reported in any document. Most likely the city opened its gates to Cyrus' troops. This shows that there were probably some prearrangements with the Babylonians; no sign of support for Nabonidus is reported. The legendary Babylon had fallen. Nabonidus was arrested in Babylon on his return (*Nabonidus Chronicle* iii: 16). In practice Nabonidus had lost his capital city Babylon years before these events.

After the army of Cyrus had entered the city, Median shield-bearers were sent to guard the gates of the Esagila Temple. There was no interruption of rituals in Esagila or the other temples of Babylon and religious ceremonies continued without break (*Nabonidus Chronicle* iii: 17–18). Cyrus' troops were forbidden to enter Babylon's sacred shrines and were kept away from the *Ekur* ('mountain house'; probably the *ziggurat* of Babylon) (*Verse Account* col. vi).

Cyrus himself entered Babylon about seventeen days later on the third day of Arahsamnu (the eighth month, 29 October). According to reports, green leaves were spread in front of him. He saluted the city and declared a state of peace. Like a Babylonian king he prayed to the gods and increased offerings to their temples. By his order, the gods that had earlier been brought by Nabonidus to Babylon were sent back to their home sanctuaries and their damaged shrines were rebuilt. Like a good Babylonian king, Cyrus commenced several projects in the city. In traditional fashion he took hoe, spade and water-basket himself and worked to complete the wall of Babylon, following the original plan of Nebuchadnezzar II. According to the *Verse Account* (col. vi), 'to the inhabitants of Babylon a joyful heart is now given. They are like prisoners when the prisons are opened. Liberty is restored to those who were surrounded by oppression. All rejoice to look upon him as king.'

Although Cyrus collected the foreign population living in captivity in Babylon and returned them to their homelands together with their gods and sacred items (Cyrus Cylinder 32), no particular people is named in the text of the Cyrus Cylinder. According to the Old Testament, Jewish captives brought by Nebuchadnezzar II to Babylon were one of those peoples to whom Cyrus returned their gold, silver and temple vessels which had been looted during the Babylonian invasion, and allowed to leave (Ezra 5:14; Josephus XI: 1.3). Other evidence confirms that different people were also sent back to their homelands at this time. The Babylonian texts discovered at Neirab in the Aleppo region in modern Syria suggest that some individuals returned to their homeland from Babylon at about this time (Hoglund 1992: 27; Fales 1973: 131–42; Eph'al 1978: 84–7). Cyrus was in fact the first ruler in history who did not simply deport people from their native lands, but rather returned them to their homelands with their treasures and sacred temple vessels.

Cyrus, then, hardly a conventional conqueror, performed the duties of a Babylonian king in different ways. For example, he carried out construction activities such as rebuilding Imgur-Enlil, the inner wall of Babylon. His troops seem to be have been well disciplined and aware of their commander's will. Whereas conquerors have generally looted treasures and destroyed the temples and holy shrines of the conquered, often carrying off their sacred symbols and treasures to their own land as victory symbols, Babylon did not witness such treatment under Cyrus, who, on the contrary, looked after the welfare of the people. He was obviously an adherent of an altogether different religion, but nevertheless respected Babylonian beliefs and those of other people who worshipped other gods.

Such a record of behaviour could be seen as false propaganda by a victorious king; however, independent texts together with other evidence confirm the message of the Cyrus Cylinder. From an archaeological point of view the excavations at Babylon, especially those of the German archaeologist Robert Koldewey, show no signs of destruction or demolition of the city at this time. Inscribed cuneiform bricks from Babylon, Ur and Uruk bearing the name of Cyrus also bear witness to his construction and reconstruction activities in those cities. Economic and legal texts from Babylon and Sippar also show that daily life continued as normal (Haerinck 1997).

One document, now in the British Museum (BM 60744), was written in 539 BC, a few days after the fall of the city and before Cyrus entered Babylon. The text is a contract recording payment of salaries in sheep. Another example, a receipt for sheep (BM 101100) from Sippar, was written six days after the capture of Sippar and four days after the fall of Babylon.

56 The stela of Nabonidus in the British Museum. The cuneiform inscription that once covered the front has been carefully erased. In the *Verse Account* Cyrus declares that 'His name was wiped out [on his monu]ments!' suggesting that the defacement of Nabonidus' works was carried out by Persian order on arrival.

BM 90837.

A further document records a land deal between certain brothers, written about a month after Cyrus entered Babylon (Strassmaier 1889: 4–5; Razmjou 2010: 52–3). Other texts dating to the post-occupation period in Babylon have similar contents.

These texts show that the legal offices and scribal houses in Babylon and Sippar operated normally under the occupation. There is no textual evidence for chaos at this time. Even by modern standards, avoiding looting and bloodshed in such circumstances can only be achieved with a highly disciplined army and the presence of an enlightened conqueror (*figure 56*).

The importance of the Cyrus Cylinder, past and present

According to the textual evidence, Cyrus was known as 'merciful' and as 'father' to the Persians (Herodotus *Histories*: 3:89), a great leader to the Greeks (Plato *Alcibiades*, 105c; Xenophon *Cyropaedia*), a liberator and peacemaker to the Babylonians and the deported peoples (*Dynastic Prophecy*; *Nabonidus Chronicle*; *Verse Account*; *Cyrus Cylinder*) and a messiah to the Jews (Isaiah 45:1). These sources present Cyrus as an extraordinarily respected figure in different cultures throughout history, in both religious and secular texts. Such a reputation makes the Cyrus Cylinder even more significant.

As Cyrus has been called 'Messiah' in the Old Testament, he also came to acquire a prophetic personality. Thus, to some people, the Cylinder containing the words of Cyrus holds a religious significance. It has been seen as confirming the account of the release of the Jews by Cyrus in the Old Testament. Besides, some sentences in the two texts, the Cyrus Cylinder and the Old Testament, are very similar; for example, in both, Marduk and Jehovah call Cyrus by name, hold his hand and walk beside him, call him 'shepherd' and lead him to capture Babylon. To others the Cylinder is also considered a symbol of religious tolerance, because it declares a respectful treatment of different gods and beliefs, Babylonian and non-Babylonian. The text shows that Cyrus respected not only the cult of the Babylonian god Marduk, but local cults as well. Unlike Nabonidus in his promotion of the moon god Sin, he was unwilling to impose his own belief, an Iranian religion, on others. It seems that Cyrus even respected the god of his opponent, Nabonidus. A cylinder fragment from Ur (UET I, no. 307), probably related to Cyrus because of its content (although this is not certain), is written for Sin, the patron god of Ur.[17]

An important aspect of the cylinder is political. The Cyrus Cylinder is neither a chronicle nor simply an account of events. There are two forms of message wrapped in its text. The first part of the text demonstrates the divine approval of Cyrus by Marduk, probably composed by Babylonian priests, while the second quotes his own words addressed to Babylon, introducing himself and reporting what he has done since the capture of the city. First of all, it was a direct message to a Babylonian audience, declaring that the old regime, with its inadequacies, is gone and that there is now a different and more beneficial system. In addition, the message was also addressed to coming generations who might find the cylinder in the wall, and become aware of his message, just as he had found that of Ashurbanipal in the same wall at Babylon. The use of traditional-style Babylonian language also helped to make the message more acceptable to the Babylonians. The Cylinder's message does not seem to be mainly concerned with legitimising Cyrus as the ruler of Babylon; this had only been urgent at the moment of conquest and was now unnecessary. The text was written later, most probably the following year, when some of the building work mentioned in the text had already been accomplished and Cyrus was long established as the legitimate ruler, the 'king of Babylon'. The text was, as we now understand, not limited to a cylinder buried as a traditional Babylonian foundation deposit; it also existed on conventional, 'above ground' clay tablets. The evidence for this is the two newly identified fragments in the British Museum collection that belong to such a text (see Chapter 1). The Cyrus Cylinder text is also carefully written in a way to preserve the pride of the Babylonian people, instead of the victor humiliating them. Different aspects of the message show that it is a well-thought-through and finely composed text encapsulating powerful political messages.

Although the Cylinder was written in Babylon in Babylonian style, both in shape and language, there are aspects in which it differs from contemporary foundation cylinders. Some may see it as propaganda by a victor and effectively indistinguishable from those of the preceding kings of Mesopotamia. However, the content and tone used in the Cyrus Cylinder are fundamentally different, even though it is written in traditional Babylonian style. As can be seen, the main focus of the text and other sources related to Cyrus is to emphasise his intention to 'enable the lands to dwell in peace'.[18] Unusually, the people are more prominent in the Cyrus Cylinder and related texts than are military successes. There is no reference to fighting, military victory or defeat in the cylinder. Rather, we hear how Cyrus' troops marched peacefully into Babylon, how the lands of Sumer and Akkad had nothing to fear, how he sought the safety of the city and its sanctuaries (lines 24–6), how he increased the offerings to temples (lines 37–8), and of course how he returned the gods and captives to their homelands (line 32).

Because of its respectful tone and the humane treatment of deportees, some have classified the Cyrus Cylinder as a 'charter of human rights'. In fact the cylinder, which is a report of the capture of Babylon and Cyrus' treatment of the city, cannot be understood as a charter in the modern sense and there is no reference to human rights as such in the text. 'Human rights' is a modern term and was not in use when the cylinder was written. There is, however, evidence of some awareness that humane treatment of people according to their 'natural rights' was the correct way, even if the concept was never articulated. This includes avoiding violence and protecting people's lives regardless of their nationality, respecting their freedom to worship gods of their choosing, to live in their homeland and the right 'to dwell in peace' – all of which today embody the natural rights of people. All these ideas are reflected in the text, and can be understood as part of a civilised and humane treatment of the conquered people of Babylon. It is noteworthy how Cyrus treated the people who were not his own people and had a different culture, language and religion. Even after the conquest of Babylon, the dignity of its people was preserved.

A comparison of the Cyrus Cylinder with the inscriptions of previous conquerors of Babylon highlights this sharply. When Sennacherib, king of Assyria (705–681 BC) captured the city in 690 BC after a 15-month siege, Babylon endured dreadful destruction and massacre. Sennacherib describes how, having captured the king of Babylon, he had him tied up in the middle of the city of Nineveh like a pig (Luckenbill 1924: 90). Then he describes how he destroyed Babylon like a storm, filled the city with corpses, looted its wealth, broke its gods, burned and destroyed its houses down to their foundations, demolished its walls and temples and dumped them in the canals. Then he flooded the city by diverting the river. After putting the inhabitants to the sword he removed the rubble and deposited it in the river Euphrates, by which it was carried away as far as the Persian Gulf. Finally, he presented samples of the very dust of Babylon to other peoples and to the temple of the New Year's Feast in Assyria (Luckenbill 1924: 83–4, 137). This was common treatment for a defeated people at this time. Sennacherib's tone in this inscription, reflecting his relish of and pride in massacre and destruction, is totally at odds with the message of the Cyrus Cylinder.

On the other hand, certain concepts in the Cyrus Cylinder are to be found in the documents of previous Babylonian and Assyrian rulers and underline the sense of continuity that was also important. A few former kings made claims close to those of Cyrus, such as Marduk-apla-iddina II (reigned 722–710 and 703–702 BC), Sargon II (722–705 BC) and Ashurbanipal (685–c. 627 BC). The last engaged in reconstruction activity in Babylon; as Cyrus says in his Cylinder and is discussed above, during reconstruction of a wall in Babylon a foundation inscription from Ashurbanipal was uncovered. Despite the familiar elements, there are obvious differences as well. Marduk-apla-iddina II, who claims he had repaired some shrines, was king of Babylon himself. Sargon's treatment of Babylon is mentioned in between reports of his burning and destruction of cities (Luckenbill 1924: 35–6, nos 68–70), and Ashurbanipal was a conqueror who, after the capture of Babylon, ordered all those who had resisted him to be killed. These are deliberately composed texts designed to threaten subject people and enemies and render them into obedient subservience. Comparing them with the case of Cyrus, the differences outweigh the similarities. The interesting point is that the author of the Cylinder text, who was familiar with previous royal literature, wisely used parts of those texts to ensure continuity but in a new way, by carefully selecting the words and content.

An other difference between the previously mentioned texts and the Cyrus Cylinder is that no other king ever returned captives to their homeland as Cyrus did. The Assyrians sometimes gave limited religious freedom to local cults and the people they conquered, but after a military conquest the conquered people usually had to submit to the 'exalted might' of the Assyrian god Ashur, their own shrines and gods were demolished and the people put under 'the yoke of Ashur'. Even Babylon itself did not show tolerance towards other beliefs and cults, for it had recently destroyed the temple in Jerusalem as well as the temple in Harran; furthermore, Nabonidus took other gods from their sacred shrines and carried them to Babylon.

In the modern political context, the cylinder has been viewed in a different way still. After the Cyrus Cylinder was briefly exhibited in Tehran in 1971 during the celebrations for the 2,500th anniversary of the founding of the Persian Empire, it became a subject for discussion. Some individuals linked it with modern politics and viewed it as an artefact that could be used by politicians for political purposes. Consequently, the Cylinder became politicised, a process which overshadowed its cultural and historical significance. Because of this, some have ignored the cultural relevance of the document and have tried to condemn it as nothing but propaganda, similar to that of other Babylonian kings before Cyrus. The Cylinder became the subject of politically oriented discussions and was thus devalued. Some even tried to interpret it from a twenty-first-century perspective, even though many concepts differ radically from those of ancient times. For example, it has been suggested that bringing tribute and kissing the feet of Cyrus by the people, as mentioned in the text, is disrespectful of those people, notwithstanding the fact that the event took place twenty-five centuries ago. Then there were different concepts, values and rituals. Bringing tribute for rulers was routine and normal practice until recent times in many cultures, regardless of whether rulers were good or bad. It is also noteworthy that even today in many countries, mainly eastern, the tradition of kissing the feet of the elders of the family, tribe or village is a highly respectful act and is never considered a sign of humiliation. In spite of these objections, many people today view the Cyrus Cylinder not as propaganda but rather as a powerful and timely cultural symbol embodying tolerance and respect for other peoples.

The Cyrus Cylinder has become both an important national icon and a symbol of cultural identity. It has different messages for different people. Some see it as a significant historical document; some see it in a religious context and as confirmation of the Old Testament; some connect it with modern politics; and yet others regard it as the first charter of human rights. It can be all or a mixture of these things.

NOTES

1. The punishment mentioned by Herodotus regarding the killing of Harpagus' son and serving him up as a meal to his father (*Histories* 1.119) is contrary to ancient Iranian beliefs and seems to be pure fiction.
2. This story is particularly close to a legend in ancient Iranian mythology described in the Avesta, the Zoroastrian holy book, and the *Shahnameh*, the Book of Kings. In the *Shahnameh*, an evil king Zahhak (Azhidahak) (with a notable similarity to the name Astyag/Astyages), also known as 'snake-shouldered' Zahhak because of two black snakes on his shoulders, ruled by tyranny. After a dream and a warning by his dream-interpreters, he ordered infants to be killed to save his throne from the newly born hero Fereidoun. The latter, also raised by a cowherd, when he grew up

defeated 'snake-shouldered' Zahhak and took the throne. These stories may have had similar mythological roots.

3. In the Sumerian story of *Enmerkar and the Lord of Aratta*, the envoy passes Anshan several times on his way to Aratta (see Cohen 1973: ll. 75, 110, 166).
4. The Persian Darius called himself 'Aryan, having Aryan lineage' (DNa: 14–15; Kent 1953: 137–8). The Persians seem to have migrated from northern Central Asia.
5. The title of king of Anshan is only used in later texts such as the *Dynastic Prophecy*, where it refers to former kings of Anshan.
6. In Babylon before Cyrus, Belshazzar, the eldest son of Nabonidus, seems to have had the same role. In Iran, such tradition continued to the Qajar dynasty, when the crown prince had to rule in Tabriz before becoming king.
7. Xenophon *Oeconomicus* 4.20–22 (later reproduced by Cicero in *On Old Age*) describes the visit of the Spartan general Lysander, who was astonished by the beauty and regularity of trees and plants in the paradise of Prince Cyrus the Younger, son of Darius II, at Sardis, designed and planted by Prince Cyrus himself.
8. Strabo *Geography* 15.3.7; Certain Greek authors such as Herodotus tell a story that Cyrus and his army were all killed in a campaign against northern tribes, while Xenophon describes how Cyrus died in his capital and was buried there. The story in Herodotus, as he heard it related in the next century, seems to be fictitious.
9. Achaemenian, or Achaemenid, is the name of the Persian dynasty, eponymously derived from Achaemenes, a legendary king who lived about 700 BC and was said to be the founder of the Persian dynasty.
10. According to Plutarch (*Alex* 69.4) the inscription read 'O man, whosoever thou art and from whencesoever thou comest, for I know thou wilt come, I am Cyrus, the founder of the Persian empire; do not grudge me this little earth which covers my body.'
11. The story in Herodotus (*Histories*: 1.86–87) of the attempt to burn Croesus alive, his life being saved by Apollo, is obviously fiction, especially since burning bodies in Iranian belief was regarded as a crime against the sacred elements. The scene has been depicted on an Attic red-figure amphora from about 500–490 BC, found at Vulci (now in the Louvre). The artist, named Myson, shows Croesus on the pyre, holding his sceptre and freely pouring a libation or a flammable liquid, while the servant Eutymos is lighting the wood with torches. This may refer to another version of the story and the fact that Croesus might have willingly decided to burn himself after his defeat, but was saved.
12. According to the Bisutun inscription of Darius, copies of his inscription were sent to other provinces in the empire to be read for the people (Kent 1953: 132).
13. In the Zoroastrian written tradition, for example in the *Dinkard*, there are references to Alexander burning the inscriptions kept at Dež-Nebešt, the 'Fortress of Archives' or the 'Castle of Inscriptions'; Shaki 1995: 348–50.
14. *Dynastic Prophecy* ii: 11–24; Grayson 1975b: 33; the translation of the last-quoted sentence has been amended by the author and the editor.
15. The older translations of this passage are problematic and dubious. For a corrected and updated translation of this passage, see Lambert 2007.
16. Also mentioned as 'Ugbaru' in the same text; he is Gaubaruva in Old Persian, Gobryas in Greek.
17. The text mentions returning gods to their shrines; Kuhrt 2007: vol. 1, 110 n. 12. In view of this and other details, it probably refers to Cyrus.
18. This view of Cyrus is also reflected in Greek works. Even Aeschylus in his play *Persians* (769), despite his contempt for Xerxes, could not hide his respect for Cyrus and refers to his good qualities: 'Cyrus, blessed in good fortune, came to the throne and established peace for all his people.'

Afterword

IRVING FINKEL

Despite the fact that a second copy of its actual wording has now come to light, the Cyrus Cylinder retains its uniqueness among objects from antiquity on other grounds. It is possessed, we may say, of two identities. The first is the original document that survives from the sixth century BC, translated with the benefit of modern understanding of Babylonian cuneiform, and to a large extent understood in its contemporary world. Its alter ego is something altogether distinct, a matter of symbolic power and embodiment of ideals that has grown beyond the literal text to become a source of vibrant inspiration in our modern world. Shirin Ebadi, accepting the Nobel Peace Prize in December 2003, declared:

> I am an Iranian. A descendant of Cyrus The Great. The very emperor who proclaimed at the pinnacle of power 2500 years ago that '… he would not reign over the people if they did not wish it'. And [he] promised not to force any person to change his religion and faith and guaranteed freedom for all. The Charter of Cyrus The Great is one of the most important documents that should be studied in the history of human rights.

Thus has Cyrus, king of Persia, come to stand clear above other kings, and his cylinder, once a localized statement of political nicety, has rolled out across the world, clad in the wish for freedom and tolerance, drawing out truth and harmony in its wake.

APPENDIX

Transliteration of the Cyrus Cylinder text

IRVING FINKEL

THE following transliteration shows the reader exactly how the ancient Babylonian text is written out in the original. Signs given in italic (such as *kiš-šat*) are syllabic signs used to spell parts of the individual words. Those in capital letters (such as LUGAL) are in the older Sumerian language and the reader has to supply the Babylonian equivalent as s/he reads, much as we supply 'dollar' when we encounter the sign $.

Sources

A BM 90920 (1880–0617.1941) + NBC 2504; lines 1–45
   Approximately two-thirds of the original cylinder; every line ruled.
   Drawing of BM cuneiform text: T.G. Pinches in Rawlinson and Pinches 1880 (cuneiform type); derivative drawing from this in Abel and Winckler 1890: 44–5.
   Drawing of NBC cuneiform text: Nies and Keiser 1920: pl. 21
   Translations: Oppenheim 1969; Schaudig 2001; Michalowski 2006
B_1 BM 47134 (1881,0830.656); lines A1–2; A42–45
B_2 BM 47176 (1881,0830.698); lines A34–37
   Two non-joining and widely separated fragments from one large tablet. Fine quality literary Late Babylonian script; every line ruled. One-line colophon. Published here for the first time.

As already mentioned, it is probable that one line of text in the Cylinder, source A, equates two lines of text in the Tablet, source B. This assumption has been made in aligning the two sources for the beginning of the inscription. The situation is less clear for the final lines. Certain signs on the Cyrus Cylinder that were copied as complete in earlier times are today less well preserved. In addition, there are erasures and other characteristics. Full details have been given in the careful edition by the scholar Hanspeter Schaudig (2001) and are not repeated here, although each sign of this inscription has been checked afresh on the Cylinder for this transliteration.

Transliteration

A1 [..]
x-ni-šu

B1,1 [...... ᵈAMAR.UT]U? LUGAL kiš-šat AN-e u KI-tì x [.........................
...................................]

B1₂ [.........šá ina] x-si-šú ú-nam-mu-[ú ....]

A2 [... ḫa-a-a-iṭ(?)
ki]-ib-ra-a-tì

B1₃ [............... ra-pa-á]š(?) uz-nam x x (x) [....................................]

Composite reconstruction of A lines 1–2:

A1 [i-nu ᵈAMAR.UT]U? LUGAL kiš-šat AN-e u KI-tì x[...šá ina] x-si-šú
ú-nam-mu-[ú (...)]x-ni-šu

A2 [....................ra-pa-á]š(?) uz-nam x x (x) [................ḫa-a-a-iṭ(?)
ki]-ib-ra-a-tì

A3 [..
.... ṣi-it] ⸢lib-bi-šu⸣ GAL ma-ṭu-ú iš-šak-na a-na e-nu-tu ma-ti-šú

4 x[.. ta]m-⸢ši⸣-li ú-ša-áš-ki-na
ṣe-ru-šu-un

5 ta-am-ši-li é-sag-íl i-te-[pu-uš-ma ....
....-t]ì a-na ŠEŠ.AB
KI ù si-it-ta-tì ma-ḫa-za

6 pa-ra-aṣ la si-ma-ti-šu-nu ta-[ak-li-im la me-si ....
...........................la] pa-liḫ ú-mi-ša-am-ma
id-de-né-eb-bu-ub ù ⸢a-na ma-ag⸣-ri-tì

7 sat-tuk-ku ú-šab-ṭi-li ú-l[a-ap-pi-it pél-lu-de-e ....
...................iš]-tak-ka-an qé-reb ma-ha-zi pa-la-ḫa ᵈAMAR.UTU
LUGAL DINGIR.MEŠ i[g-m]ur kar-šu-uš-šu

8 le-mu-ut-ti URU-šu [i-t]e-né-ep-pu-⸢uš⸣ u₄-mi-ša-am-⸢ma⸣ x x

[... UN].MEŠ-šú i-na ab-ša-a-ni la ta-ap-šu-úḫ-tì ú-ḫal-li-iq kul-lat-si-in

9 a-na ta-zi-im-ti-ši-na ᵈen-líl DINGIR.MEŠ ez-zi-iš i-gu-ug-m[a ....
.............................] ki-su-úr-šu-un DINGIR.MEŠ a-ši-ib lib-bi-šu-nu i-zi-bu at-ʳmaʳ-an-šu-un

10 i-na ug-ga-ti-ša ú-še-ri-bi a-na qé-reb šu-an-na KI ᵈAMAR.UTU t[i-iz-qa-ru ᵈen-líl DINGIR.M]EŠ us-sa-aḫ-ra a-na nap-ḫar da-ád-mi ša in-na-du-ú šu-bat-su-un

11 ù UN.MEŠ KUR šu-me-ri ù URI KI ša i-mu-ú ša-lam-ta-áš ú-sa-ʳaḫʳ-ḫi-ir ka-ʳbatʳ-[ta-áš] ir-ta-ši ta-a-a-ra kul-lat ma-ta-a-ta ka-li-ši-na i-ḫi-iṭ ib-re-e-ma

12 iš-te-'e-e-ma ma-al-ki i-šá-ru bi-bil lib-bi-ša it-ta-ma-aḫ qa-tu-uš-šu ᵐku-ra-áš LUGAL URU an-ša-an it-ta-bi ni-bi-it-su a-na ma-li-ku-tì kul-lat nap-ḫar iz-zak-ra šu-ʳum-šúʳ

13 KUR qu-ti-i gi-mir um-man-man-da ú-kan-ni-ša a-na še-pi-šu UN.MEŠ ṣal-mat SAG.DU ú-šak-ši-du qa-ta-a-šú

14 i-na ki-it-tì ù mi-šá-ru iš-te-né-'e-ši-na-tì ᵈAMAR.UTU EN GAL ta-ru-ú UN.MEŠ-šú ep-še-ti-ša ù dam-qa-a-ta ù lib-ba-šu i-ša-ra ḫa-di-iš ip-pa-li-i[s]

15 ana URU-šú KÁ.DINGIR.MEŠ KI a-la-ak-šu iq-bi ú-ša-aṣ-bi-it-su-ma ḫar-ra-nu TIN.TIR KI ki-ma ib-ri ù tap-pe-e it-tal-la-ka i-da-a-šu

16 um-ma-ni-šu rap-ša-a-tì ša ki-ma me-e ÍD la ú-ta-ad-du-ú ni-ba-šu-un GIŠ.TUKUL.MEŠ-šu-nu ṣa-an-du-ma i-ša-ad-di-ḫa i-da-a-šu

17 ba-lu qab-li ù ta-ḫa-zi ú-še-ri-ba-áš qé-reb šu-an-na KI URU-šu KÁ.DINGIR.MEŠ KI i-ṭi-ir i-na šap-ša-qí ᵐᵈAG-NÍ.TUKU LUGAL la pa-li-ḫi-šu ú-ma-al-la-a qa-tu-uš-šu

18 UN.MEŠ TIN.TIR KI ka-li-šu-nu nap-ḫar KUR šu-me-ri u URI KI ru-bé-e ù šak-ka-nak-ka ša-pal-šu ik-mi-sa ú-na-áš-ši-qu še-pu-uš-šu iḫ-du-ú a-na LUGAL-ú-ti-šú im-mi-ru pa-nu-uš-šú-un

19 be-lu ša i-na tu-kul-ti-ša ú-bal-li-ṭu mi-tu-ta-an i-na pu-uš-qu ù ú-de-e ig-mi-lu kul-la-ta-an ṭa-bi-iš ik-ta-ar-ra-bu-šu iš-tam-ma-ru zi-ki-ir-šu

20 a-na-ku ᵐku-ra-áš LUGAL kiš-šat LUGAL GAL LUGAL dan-nu LUGAL TIN.TIR KI LUGAL KUR šu-me-ri ù ak-ka-di-i LUGAL kib-ra-a-ti er-bé-et-tì

21 DUMU ᵐka-am-bu-zi-ia LUGAL GAL LUGAL URU an-ša-an DUMU DUMU ᵐku-ra-áš LUGAL GAL LUGA[L U]RU an-ša-an ŠÀ.BAL.BAL ᵐši-iš-pi-iš LUGAL GAL LUGAL URU an-šá-an

22 NUMUN da-ru-ú ša LUGAL-ú-tu ša ᵈEN u ᵈAG ir-a-mu pa-la-a-šu a-na ṭu-ub lib-bi-šú-nu iḫ-ši-ḫa L[UGA]L-ut-su e-nu-ma an[a q]é-reb TIN.TIR KI e-ru-bu sa-li-mi-iš

23 i-na ul-ṣi ù ri-ša-a-tì i-na É.GAL ma-al-ki ar-ma-a šu-bat be-lu-tì ᵈAMAR.UTU EN GAL lib-bi ri-it-pa-šu ša ra-ʳimʳ TIN.TIR KI ši-m[a]-ʳa-tišʳ iš-ku²-ʳnaʳ-an-ni-ma u₄-mi-šam a-še-'-a pa-la-ʳaḫʳ-šú

24 um-ma-ni-ia rap-ša-a-tì i-na qé-reb TIN.TIR KI i-ša-ad-di-ḫa šu-ul-ma-niš nap-ḫar KU[R šu-me-ri] ʳùʳ URI KI mu-gal-[l]i-tì ul ú-šar-ši

25 ⸢URU KI⸣ KÁ.DINGIR.RA KI ù kul-lat ma-ḫa-zi-šu i-na ša-li-im-tì
áš-te-'e-e DUMU.MEŠ TIN.TIR [KI...š]a ki-ma la lib-[bi DING]IR-ma
ab-šá-a-ni la si-ma-ti-šú-nu šu-ziz-⸢zu⸣

26 an-ḫu-ut-su-un ú-pa-áš-ši-ḫa ú-ša-ap-ṭi-ir sa-ar-ma-šu-nu a-na ep-še-ti-[ia
dam-qa-a-ti] ᵈAMAR.UTU EN GA[L]-ú iḫ-de-e-ma

27 a-na ia-a-ti ᵐku-ra-áš LUGAL pa-li-iḫ-šu ù ᵐka-am-bu-zi-ia DUMU ṣi-it
lib-bi-[ia ù a-n]a nap-ḫ[ar] um-ma-ni-ia

28 da-am-qí-iš ik-ru-ub-ma i-na šá-lim-tì ma-ḫar-ša ṭa-bi-iš ni-it-t[a-al-la-ak i-na
qí-bi-ti-šú] ṣir-ti nap-ḫar LUGAL a-ši-ib um-ma-ni-ia BÁRA.MEŠ

29 ša ka-li-iš kib-ra-a-ta iš-tu tam-tì e-li-tì a-di tam-tì šap-li-tì a-ši-ib n[a-gi-i
né-su-tì] LUGAL.MEŠ KUR a-mur-ri-i a-ši-ib kuš-ta-ri ka-li-šú-un

30 bi-lat-su-un ka-bi-it-tì ú-bi-lu-nim-ma qé-er-ba šu-an-na KI ú-na-áš-ši-qu
še-pu-ú-a iš-tu [šu-an-na K]I a-di URU aš-šur KI ù MÙŠ.EREN KI

31 a-kà-dè KI KUR èš-nu-nak URU za-am-ba-an URU me-túr-nu BÀD.
DINGIR KI a-di pa-aṭ KUR qu-ti-i ma-ḫa-z[a e-be]r-ti ÍD.IDIGNA ša iš-tu
pa-na-ma na-du-ú šu-bat-su-un

A32 DINGIR.MEŠ a-ši-ib lìb-bi-šú-nu a-na áš-ri-šu-nu ú-tir-ma ú-šar-ma-a
šu-bat da-rí-a-ta kul-lat UN.MEŠ-šú-nu ú-pa-aḫ-ḫi-ra-am-ma ú-te-er
da-ád-mi-šú-un

A33 ù DINGIR.MEŠ KUR šu-me-ri ù URI KI ša ᵐᵈAG-NÍ.TUKU a-na
ug-ga-tì EN DINGIR.MEŠ ú-še-ri-bi a-na qé-reb šu-an-na KI i-na qí-bi-ti
ᵈAMAR.UTU EN GAL i-na ša-li-im-tì

B21' (unplaced traces)

B22' [.. a-n]a u[g-ga]-tì EN
DINGIR.MEŠ ú-še-ri-⸢bi⸣ [...........]

A34 i-na maš-ta-ki-šu-nu ú-še-ši-ib šu-ba-at ṭu-ub lib-bi kul-la-ta DINGIR.MEŠ
ša ú-še-ri-bi a-na qé-er-bi ma-ḫa-zi-šu-un

B23' [................................. ṭu-u]b lib-bi kul-lat DINGIR.MEŠ šá
ú-še-r[i-bi ....]

A35 u₄-mi-ša-am ma-ḫar ᵈEN ù ᵈAG ša a-ra-ku UD.MEŠ-ia li-ta-mu-ú lit-
taz-ka-ru a-ma-a-ta du-un-qí-ia ù a-na ᵈAMAR.UTU EN-ia li-iq-bu-ú ša
ᵐku-ra-áš LUGAL pa-li-ḫi-ka u ᵐka-am-bu-zi-ia DUMU-šú

B24' [.............................. U]D.MEŠ-ia li-ta-mu-ú lit-taz-ka-ru a-
[..............................]

A36 ku x [x x x-i]b? šu-nu lu ú-⸢za-ni-ni⸣ (illegible traces) ù(?) UN.MEŠ TIN.
TIR KI ⸢ik-tar-ra-bu⸣ LUGAL-ú-tu KUR.KUR ka-li-ši-na šu-ub-ti né-eḫ-tì
ú-še-ši-ib

B25' [......... x ⸢šu-nu⸣ lu-ú za-ni-ni BÁRA-i-ni a-na ⸢UD.MEŠ⸣ S[ÚD.MEŠ(?)
....]

37 [...................................... KUR].GI.MUŠEN
2 UZ.TUR.MUŠEN ù 10 TU.GUR₄.MUŠEN.MEŠ e-li KUR.
GI.MUŠEN UZ.TUR.MUŠEN.MEŠ ù TU.GUR₄.MUŠEN.MEŠ

B₂6′ (unplaced traces)

38 [.. u₄-m]i-šam ú-ṭa-aḫ-ḫi-id BÀD im-gur-ᵈen-líl BÀD GAL-a ša TIN.TIR K[I ma-aṣ-ṣ]ar-ʿtaʾ-šú du-un-nu-nù áš-teʾe-ma

39 [...............................] ka-a-ri a-gur-ru šá GÚ ḫa-ri-ṣi ša LUGAL maḫ-ri i-p[u-šu-ma la ú-ša]k-ʿli-luʾ ši-pi-ir-šú

40 [........................... la ú-ša-as-ḫi-ru URU ʿaʾ-na ki-da-a-ni ša LUGAL ma-aḫ-ra la i-pu-šu um-man-ni-šu di-ku-u[t ma-ti-šu i-na/a-na q]é-ʿrebʾ šu-an-na KI

41 [..................... i-na ESIR.ḪÁD.RÁ]ʿAʾ ù SIG₄.AL.ÙR.RA eš-ši-iš e-pu-uš-ma [ú-šak-lil ši-pir-ši]-in

A42 [........................ GIŠ.IG.MEŠ GIŠ.EREN MAḪ].MEŠ ta-aḫ-lu-up-tì ZABAR as-ku-up-pu ù nu-ku-š[e-e pi-ti-iq e-ri-i e-ma KÁ.MEŠ-š]i-na

B₁1′ (unplaced traces: could prove to belong to A42 or 43)

A43 [ú-ra-at-ti .... š]i-ṭi-ir šu-mu šá ᵐAN.ŠÁR-DÙ-IBILA LUGAL a-lik maḫ-ri-[ia šá qer-ba-šu ap-pa-a]l-sa

B₁2′ [.........................] ᵐAN.ŠÁR-DÙ-ʿAʾ [........................ ....]

A44 [... ....] x x x [x x x]-x-tì

B₁3′ [............a-na áš-r]i-ʿšúʾ ᵈAMAR.UTU EN GAL ba-l[aṭ u₄-um re-e-qú-ú-ti]

A45 [... .... a-na d]a-rí-a-tì

B₁4′ [še-bé-e li-it-tu-ú-ti ku-un GIŠ.GU.ZA ù la-bar pa-le]-ʿeʾ a-na ši-ri-ik-t[i šu-úr-kam]

B₁5′ [ù a-na-ku-ma .... li]b-bi-ka a-na da-[rí-a-tì]

Colophon in B₁:

B₁6′: [ki KA … šá-ṭir b]a-ar IM ᵐNÍG.BA-ᵈAMAR.UTU ʿAʾ [......] [Written and check]ed [from a…]; (this) tablet (is) of Qīšti-Marduk, son of […].

Name: Qīšti-Marduk or Iqīš-Marduk

APPENDIX 133

Notes on the inscription

Explanations of the Babylonian words and forms in the Cyrus Cylinder have been given by P.-R. Berger, and more recently Hanspeter Schaudig, and the interested reader is best referred to both works (Berger 1975 and Schaudig 2001). Some remarks raised by the identification of the two duplicating fragments follow, while we now have a better understanding of the very end of the text, the restoration of which owes much to the extensive knowledge of Professor Berger of Babylonian royal inscriptions.

1–2. It is still not certain how the surviving phrases at the beginning of the Cylinder A and the new fragment B₁ should be put together. The final word in A line 2, *kibrāti*, 'world regions', suggests that both lines 1 and 2 of the cylinder were given over to praising Marduk, as seems altogether natural, and here the partial line 1 of B₁ fits perfectly. Line 2 in B₁, however, contains only the broken phrase 'the … who, in] his …, lays waste the …' This verb of destruction hardly belongs among laudatory phrases addressed to Marduk, although a tempting restoration is 'who lays waste the shrines', which would fit with the position adopted against Nabonidus that is clearly the subject of lines 3 and beyond, while the expression actually occurs in an inscription of Nabonidus himself. The problem is that this phrase in B should come well before the *kibrāti* at the end of A line 2, and thus seems out of place. Given the understanding that one line in the Cylinder A represents two or three in B it is impossible to figure that B line 2 belongs in the gap at the beginning of A line 3, where it would make excellent sense.

3. Attention is drawn to the new reading of the signs ṣi-it] ⸢lib-bi-šu⸣ GAL, 'his [first]born', which is important as it confirms that the reference is to Belshazzar, the son of Nabonidus, whose activities as 'supply ruler' proved unacceptable to the Babylonians and earned him, like his father, the condemnation as a 'counterfeit'.

36. The new reading provided by source B of much of the beginning of this line is of substantial importance for our understanding of the nature and background to the whole cylinder. The phrase 'May they [said of Cyrus and Cambyses] be the provisioners of our temples' is surely an allusion to,

or rather an adjusted quotation from, Tablet V of the Babylonian Creation Epic, where it appears in the following context:

iš-tu u₄-mi šu-ú lu-ú za-ni-nu BÁRA-*i-ni*
'From this day on let him be the provisioner of our shrine'

The announcement of the sovereignty of Marduk as king of the gods in the Creation Epic requires him to acknowledge the upkeep of the temples of the other gods as his first responsibility. The allusion in the present context can be no coincidence: the status of Cyrus as king and Cambyses (by implication) after him requires acknowledgement by both – as is shown by the substitution of the plural personal pronoun (*šunu*) that they will similarly guarantee the maintenance of the newly rehoused gods of Babylonia. This must reflect the deliberate intention to ally the role of Cyrus with that of Marduk and is a remarkable instance of theological and political evolution in process. A similar case occurs with line 17, 'he saved his city Babylon from hardship', with a line from Tablet VI of the Babylonian Creation Epic in which Marduk builds Babylon (Schaudig 2001: 555 n906). The text now throws a more intelligible light on the subsequent passage in which details of cult offerings are laid out in detail, which, prior to the new source B, have been rather a mysterious component in the Cylinder's text, especially when it is considered to operate on an altogether different level. Now it is clear that the gods are laying out the specifics of what they expect for their upkeep, and even demanding a higher level of service than prevailed before.

References

Abel, L., and H. Winckler (1890) *Keilschrifttexte zum Gebrauch bei Vorlesungen*. Berlin.
Anonymous (1962) 'Of all the museums in the world...', *Display*, September: 41.
Ardakani, A., and N. MacGregor (2010) *The Cyrus Cylinder, 12 September 2011–12 January 2011, National Museum of Iran*. Tehran.
Bailey, M. (2004) 'How Britain Tried to Use a Persian Antiquity for Political Gain', *Art Newspaper* 150 (1 September).
Berger, P.-R. (1975) 'Der Kyros-Zylinder mit dem Zusatzfragment BIN II Ni. 32 und die akkadischen Personennamen im Danielbuch', *Zeitschrift für Assyriologie* 64: 192–234.
Briant, P. (2002) *From Cyrus to Alexander: A History of the Persian Empire*, trans. P.T. Daniels. Winona Lake, IN.
British Museum (1922) *Guide to the Babylonian and Assyrian Antiquities*. London.
——— (1931) *Guide to an Exhibition of Persian Art, in the Prints and Drawings Gallery*. London.
——— (1932) *Summary Guide to the Exhibition Galleries of the British Museum*. London.
——— (1952) *Summary Guide to the Antiquities of Western Asia*. London.
——— (1963) *Catalogue: Plaster Casts*. London.
——— (1992) *A Catalogue of Casts*. London.
Budge, E.A.W. (1880) *The History of Esarhaddon, Son of Sennacherib, King of Assyria, B.C. 681–68*. London.
——— (1884) *Babylonian Life and History*, 1st edn. London.
——— (1920) *By Nile and Tigris: A Narrative of Journeys in Egypt and Mesopotamia on Behalf of the British Museum between the Years 1886 and 1913*. London.
——— (1925) *Babylonian Life and History*, 2nd edn. London.
Caygill, M. (1981) *The Story of the British Museum*. London.
Clancier, P. (2009) *Les bibliothèques en Babylonie dans la deuxième moitié du 1er millénaire av. J.-C.* Münster.
Cohen, S. (1973) *Enmerkar and the Lord of Aratta*. Michigan: University Microfilms International.
Curtis, J., and N. Tallis (eds) (2005), *Forgotten Empire: The World of Ancient Persia*. London.

Curtis, V.S. (2011) 'Fascination with the Past; Ancient Persia on the Coins and Banknotes of Iran', in S. Bhandare and S. Garg (eds), *Felicitas: Essays in Numismatics, Epigraphy and History in Honour of Joe Cribb*, pp. 80–99. Mumbai.

Da Riva, R. (2008) *The Neo-Babylonian Royal Inscriptions: An Introduction*. Münster.

Ellis, R.S. (1968) *Foundation Deposits in Ancient Mesopotamia*. Yale: Yale Near Eastern Researches 2.

Eph'al, I. (1978) 'The Western Minorities in Babylonia in the 6th–5th Centuries B.C.E.: Maintenance and Cohesion', *Orientalia* (NS) 47: 74–90.

Fales, M. (1973) *Censimenti e catasti di epoca neo-assira*. Rome: Studi economici e tecnologici 2.

Farahbakhsh, F.N. (2005) *Standard Catalogue of Iranian Banknotes 2005*. Tehran.

—— (2010) *The Stamps of Iran: Qajar, Pahlavi, Islamic Republic of Iran 2010*. Tehran.

Finkel, I.L. (2008) 'Belshazzar's Feast and the Fall of Babylon', in I.L. Finkel and M.J. Seymour (eds), *Babylon, Myth and Reality*, pp. 170–78. London.

Finkel, I.L., and M.J. Seymour (eds) (2008) *Babylon, Myth and Reality*. London.

Frame, G. (1995) *Rulers of Babylonia from the Second Dynasty of Isin to the End of Assyrian Domination (1157–612 BC)*. The Royal Inscriptions of Mesopotamia Babylonian Periods vol. 2. Toronto.

Ghias-Abadi, R.M. (2001) *Cylinder of Cyrus*, 3rd edn. Tehran.

Grayson, A.K. (1975a) *Assyrian and Babylonian Chronicles*. Locust Valley NY: Texts from Cuneiform Sources 5.

—— (1975b) *Babylonian Historical-Literary Texts*. Toronto.

Haerinck, E. (1997) 'Babylonia under Achaemenid Rule', in J. Curtis (ed.), *Mesopotamia and Iran in the Persian Period. Conquest and Imperialism 539–331 BC*: 26–34. London.

Harmatta, J. (1971) 'The Literary Patterns of the Babylonian Edict of Cyrus', *Acta Antiqua Academitae Scientiarum Hungaricae* 19/3–4: 207–31.

Hawkes, J. (1962) 'Stirring Up Dust in the British Museum', *Observer*, 17 June 1962, p. 10.

Hilprecht, H.V. (1903) *Explorations in Bible Lands during the 19th Century*. Philadelphia.

Hoglund, K.G. (1992) *Achaemenid Imperial Administration in Syria–Palestine and the Missions of Ezra and Nehemiah*. Atlanta GA: SBL dissertation Series 125.

Ismail, M. (2011) *Wallis Budge: Magic and Mummies in London and Cairo*. Kilkerran.

Kent, R.G. (1953) *Old Persian: Grammar, Texts, Lexicon*, 2nd edn. New Haven, CT.

Kuhrt, A. (1983) 'The Cyrus Cylinder and Achaemenid Imperial Policy', *Journal for the Study of the Old Testament* 25: 83–97.

—— (2007) *The Persian Empire*, 2 vols. London and New York.

Lambert, W.G. (2007) 'Cyrus' Defeat of Nabonidus', *Nouvelles Assyriologiques Brèves et Utilitaires* 4 (2007): 13–14.

Luckenbill, D.D. (1924) *The Annals of Sennacherib*. Chicago.

Michalowski, P. (2006) 'The Cyrus Cylinder', in *Historical Sources in Translation*, pp. 426–30. Oxford.

—— (forthcoming) 'Biography of a Sentence: Assurbanipal, Nabonidus, and Cyrus', in W.F.M. Henkelman, M. Kozuh and C. Woods (eds), *Extraction and Control: Festschrift for Matthew Stolper*. Chicago.

Nies, J.B., and C.E. Keiser (1920) *Historical, Religious and Economic Texts and Antiquities*. New Haven, CT.

Oppenheim, A.L. (1969) 'Babylonian and Assyrian Historical Texts', *Ancient Near Eastern Texts Relating to the Old Testament*, ed. J.B. Pritchard, 3rd edn, pp. 265–317. Princeton, NJ.

Pahlavi, Mohammad Reza Shah (1961) *Mission for My Country*. London.

——— (1967) *The White Revolution of Iran*. Tehran.
Peters, J.P. (1897) *Nippur, or, Explorations and Adventures on the Euphrates*. New York and London.
Pinder-Wilson, R. (ed.) (1971) *Royal Persia*. London.
Pritchard, J.B. (1950) *Ancient Near Eastern Texts Relating to the Old Testament*. Princeton, NJ.
Rassam, H. (1881) 'Recent Assyrian and Babylonian Research', *Journal of the Transactions of the Victoria Institute* XIV: 182–220.
——— (1897) *Asshur and the Land of Nimrod*. New York.
Rawlinson, H.C. (1846) *The Persian Cuneiform Inscription at Behistun, decyphered and translated, with a Memoir on Persian Cuneiform Inscriptions in General, and on that of Behistun in particular*. Royal Asiatic Society. London.
——— (1880) 'Notes on a Newly-discovered Clay Cylinder of Cyrus the Great', *Journal of the Royal Asiatic Society* (NS) 12: 70–97.
Rawlinson, H.C., and T.G. Pinches (1880) *The Cuneiform Inscriptions of Western Asia*, vol. 5. London.
Razmjou, Sh. (2010) *The Cylinder of Cyrus the Great* (In Persian and English). Tehran.
Reade, J.E. (1986) 'Rassam's Babylonian Collection: The Excavations and the Archives', *Catalogue of the Babylonian Tablets in the British Museum*. Volume VI: *Tablets from Sippar*, Intro. E. Leichty, pp. xiii–xxxvi. London.
——— (1993) 'Hormuzd Rassam and His Discoveries', *Iraq* 55: 39–62.
Rich, C.J. (1815) *Memoir of the Ruins of Babylon*. London (1st edn, 2nd edn 1816; 3rd edn 1818).
——— (1839) *Narrative of a Journey to the Site of Babylon in 1811. Memoir on the ruins. Remarks on the topography of ancient Babylon by Major Rennell in reference to the memoir. Second memoir on the ruins in reference to Major Rennell's remarks. With narrative of a journey to Persepolis*. London.
Sami, A. (1970) *Persepolis (Takht-i-Jamshid)*. Shiraz.
Schaudig, H. (2001) *Die Inschriften Nabonids von Babylon und Kyros' des Grossen*. Münster.
Selby, W.B. (1859) *Memoir on the Ruins of Babylon*. Selections from the Memoirs of the Bombay Government, New Series, vol. 51. Bombay.
Shaki, M. (1995) 'Dež ī Nebešt', *Encyclopaedia Iranica*, vol. VII: Dārā(b) – Ebn al-Atīr, pp. 348–50. Tehran.
Sharp, R.N. (1966) *The Inscriptions in Old Persian Cuneiform of the Achæmenian Emperors*. Tehran.
Smith, M. (1963) 'II Isaiah and the Persians', *Journal of the American Oriental Society* 83: 415–21.
Strassmaier, J.N. (1889) *Inschriften von Cyrus, König von Babylon (538–529 v. Chr.)*. Leipzig.
Stronach, D. (1965) 'Excavations at Pasargadae: Third Preliminary Report', *Iran* 3: 9–40, pls I–XIV.
——— (1978) *Pasargadae: A Report on the Excavations Conducted by the British Institute of Persian Studies from 1961 to 1963*. Oxford.
——— (1990) 'The Garden as a Political Statement: Some Case Studies from the Near East in the First Millenium B.C.', *Bulletin of the Asia Institute* 4: 171–80.
van der Spek, R.J. (2003) 'Darius III, Alexander the Great and Babylonian Scholarship', in W. Henkelman and A. Kuhrt (eds), *A Persian Perspective: Essays in Memory of Heleen Sancisi-Weerdenburg*, pp. 289–346. Leiden: Achaemenid History XIII.
——— (forthcoming) 'Cyrus the Great, Exiles and Foreign Gods: A Comparison of

Assyrian and Persian Policies on Subject Nations', in W.F.M. Henkelman, M. Kozuh and C. Woods (eds), *Extraction and Control: Festschrift for Matthew Stolper*. Chicago.

Walker, C.B.F. (1972) 'A Recently Identified Fragment of the Cyrus Cylinder', *Iran* 10: 158–9.

——— (1987) *Cuneiform*. London.

Waterfield, G. (1963) *Layard of Nineveh*. London.

Wood, B. (2000) '"A Great Symphony of Pure Form": The 1931 International Exhibition of Persian Art and Its Influence', *Ars Orientalis* 30: 113–30.

Wu, Yuhong (1986) 'A Horse-Bone Inscription copied from the Cyrus Cylinder (Line 18–21) in the Palace Museum in Beijing', *Journal of Ancient Civilizations* 1: 15–20.

Yang Zhi (1987) 'Brief Note on the Bone Cuneiform Inscriptions', *Journal of Ancient Civilizations* 2: 131–4.

Index

References to illustrations are in italics.

al-Abid, Ahmed, 42
Adda-guppi, 105
Ahmadinejad, Mahmud, 92–3, 97, *98*, 99
Akkad, 5, 6, 7, 9
Amran, 56–8
Amurru, 6, 9
Anchang, Shi, 29
Anshan, 5, 6, 9
  Cyrus II, King of, 110–11
Antiquities of the Jews (Josephus), 25
Ardakani, Azadeh, 94, 98, 100, 101
Ashur, 6, 9
Ashurbanipal, 7, 9, 11, 22, 26
Asshur and the Land of Nimrod (Rassam), 56–7
Astyages, King of Media, 108–9

Babylon
  conquest of (by Cyrus II), 1–2
  Cylinder, reference to, 6, 7
  before Cyrus, 104–8
  excavations (1879), background to, 38–40
  fall of, 116–19
  Jumjuma, location of, 58
  location of, *10*
  Hormuzd Rassam, excavations by, 42–4
  Sennacherib, capture of the city, 122
Babylonian Life and History (Budge), 32–3, *32*
Baghaei, Hamid, 94, 97, 98–9, 100–101
Bailey, Martin, 88

Baltazar, Malcolm, 48, 50
Barnett, Richard, 88
Bel, 6, 7
Belshazzar, 4–5, 9
  ruling of Babylon, 107
Berger, P.-R., 13, 15, 134
Berossus, 108
Birch, Samuel, 46–7
  Hormuzd Rassam, correspondence with, 35–6, *37*, 43, 45, 50, 59–60
Bond, Edward, correspondence with Hormuzd Rassam, 45–6
'bone texts', forgeries (Chinese), 26–34, *27*
  discovery of, 26–8
  history and publication of, 28–9
  questions raised over, 30–31
British Museum
  1881,0830 collection, Babylon, 18–20, *19*
  cuneiform tablets, 2, 40
  Cyrus Cylinder workshop (2010), 2–3
  display of Cylinder, 69–82
  gallery photographs, *70*
  Guide to the Babylonian and Assyrian Antiquities (1922), 85
  MacGregor, Neil, 91, 94, 95, 97, 98, 100, 101, 102
  Persian Landing, 74
  Persian Room, 71–5, *72*, *73*
  repair and conservation of the Cylinder, 14
  Yale fragment, loan of, 15
Budge, E.A. Wallis, 32

Cambridge Ancient History, 85
Cambyses, 6, 7
cast, of Cylinder, 80–82, *81*
Clay, Albert T., 13
cuneiform tablets in the British Museum
    duplicate text on, 2, 18–23, *19*, 134–5
    questions raised, 134–5
Cylinder, *8*, *65*
    Ashurbanipal, link to text of, 26
    'bone texts', forgeries (Chinese), 26–34, *27*
    breakage, 54–6
    British Museum workshop, 2–3, 94
    burial of, 11, 18
    cast of, 80–82, *81*
    comparable cylinders, *65*
    copy, made by Pinches, 60–62
    cuneiform, form of, 12
    Cyrus II, commemorative items, *88*, *89*
    damage to, 55–6
    differing types of cylinder, *66*
    discovery of, 36–7, 47–9, *49*–*50*
    display of (British Museum), 69–82
    findspot of, 56–9
    'flattened out' views, *16*–*17*
    as a foundation deposit, 18, 64
    gallery photographs, *70*
    as 'human rights' declaration, 86–7, 122
    inscription, 1–2, 11; *see also* translation
    Iranian Room (British Museum), *78*, *79*
    literary sections of, 23–4
    materials and manufacture, 11–12, *66*–*7*
    names mentioned, 9
    as object, 11–12, 64
    Old Testament parallels, 25–6, 120
    Persian Room display, *72*, *73*
    purpose, of the text, 24–5
    Rahim Irvani gallery, British Museum, *77*, *80*
    re-firing of, 80
    repair and conservation, at British Museum, 14
    shipment and arrival, British Museum, 52–4
    significance of, 60–62, 120–24
    summary: discovery and significance of, 62–4
    Tehran, loan to (1971), 76, 88
    Tehran, loan to (2010–11), 91–2, 93–103, *95*, *96*, *97*
    translation, in English, 4–7
    transliteration of text, 130–33
    'uniqueness' of, 15
    Yale fragment, 13–15, *14*
cylinders, types of, *66*
Cyrus Cylinder, *see* Cylinder
Cyrus II
    Anshan, King of, 110–11
    Babylon, conquering of, 1–2
    burial of the Cylinder, 11, 18
    captives, return of, 117–18, 122, 123
    commemorative items, *88*, *89*, *90*, *91*, *92*, *93*
    on Cylinder, reference to, 5, 6, 7
    fall of Babylon, 116–19
    Founding of the Persia Empire celebrations, 73–6, *86*, 87–8
    Herodotus, *Histories*, 108–9, 115, 120
    'king of kings', 111
    Lydia, conquest of, 114
    Pasargadae, 111–14
    Persia, entry into, 108–10
    tomb of, Pasargadae, *113*
    Xenophon, *Cyropaedia*, 109, 115, 120

Dalley, Stephanie, 29
Der, 6, 9
discovery, of the Cylinder, 47–9
    sources of information for, 36–7
display of Cylinder, British Museum
    gallery photographs, *70*
    Iranian Room, *78*, *79*
    Persian Art Exhibition (1931), 69
    Persian Room, 71–5, *72*, *73*
    Rahim Irvani gallery, 77, *80*
Dongting, Wang, 28
Dynastic Prophecy, 105, 108, 116, 120

East India Company, 38
Ebadi, Shirin, 127
Encyclopedia Britannica (1929), 85–6
Enlil, 5, 9
Esagil, 4, 9
    findspot of Cylinder, 59
Eshnunna, 6, 9
excavations at Babylon (1879)
    background to, 38–40
    commencement of, by Rassam, 42–3
    looting and illegal digging, 42–3
Ezra, Book of, 25

findspot, of Cylinder, 56–9
foundation deposit, 18, 64

Gotch, Paul, 75
Gray, G. Buchanan, 85
Gu, Lanpo, 28

Guide to the Babylonian and Assyrian Antiquities (British Museum), 85
Gurney, O.R., 26, 29
Guti, 5, 9

Harmatta, J., 26
Harran, 105
Herodotus, *Histories*, 108–9, 115, 120
Hilprecht, H.V., 56–8
Historical, Religious and Economic Texts and Antiquities (Nies and Keiser), 13
'human rights' declaration, 86–7, 122

ICHHTO (Iranian Cultural Heritage, Handicrafts and Tourism Organization), 94, 95, 99, 100
Imgur-Enlil, 7, 9
inscription
    on cuneiform tablets in the British Museum, 2
    on Cylinder, 1–2, 11
    Old Testament parallels, 25–6, 120
    tablets, text on, and questions, 134–5
    see also translation
Iranian Room (British Museum), *78, 79*

Jones, John Winter, 40–41, 42, 43
Josephus, *Antiquities of the Jews*, 25
Jumjuma, location of, 58

Koldewey, Robert, 118

Labashi-Marduk, 104
Lambert, W.G., 18, 21
Layard, Austen Henry, 38, 40, 47
Lin, Zhichun, 28, 29
literary origins, of Cylinder inscription, 23–4
Lockett, Captain, 38
Luo, Guoyi, 29
Luo Xuetang, 28
Lydia, conquest of, 114

Ma, Xueliang, 29
MacGregor, Neil, 91, 94, 95, 97, 98, 100, 101, 102
Map of the Ruins at Babylon (Selby), 57
Marduk, 4, 5, 6, 7, 9, 24, 25, 105
    king of the gods, 135
Mashaei, Esfandiya Rahim, 97, 101–2
materials and manufacture of Cylinder, 11–12, 66–7
Memoirs on the Ruins of Babylon (Rich), 38
Meturnu, 6, 9

Mignan, Captain Robert, 38
Miles, Samuel Barrett, 46

Nabonidus, 1, 5, 7
    fall of Babylon, 116–19
    reign of Babylonia, 104–8
    Sin, devotion to, 105–6
    stela of (British Museum), *119*
Nabonidus Chronicle, 107, 108, 109, 110, 117, 120
Nabopolassar, 1
Nabu, 6, 7, 9
Nebuchadnezzar II, 1, 104
Nies, Rev. James B., 13, 55
Nixon, Colonel J.P., 45, 46

Pahlavi, Mohammad Reza Shah, 86
    The White Revolution of Iran, 87
Palace Museum (Beijing, China), 26, 29
Pasargadae, 111–14
    Cyrus II, tomb of, *113*
Pasha, Mohammed, 42
Persian Art Exhibition, British Museum (1931), 69
Persian Room (British Museum), 71–5, *72, 73*
    Smith, Sidney, 71
Peters, J.P., 58
Pinches, Theophilus G., 32, 33, 55, 63
    copy of Cylinder, 60–62
    inventory of tablets and other finds from Babylon area, 50–52, *53*
    Rassam, Hormuzd, correspondence with, 60

Qīštī-Marduk, 7, 22–3

Rahim Irvani gallery, British Museum, 77, *80*
Rassam, Hormuzd, 12, 35, 39, 85
    Birch, Samuel, correspondence with, 35–6, *37*, 43, 45, 50, 59–60
    Bond, Edward, correspondence with, 45–6
    Esagil, excavations of (1880), 59
    excavations and background to (Babylon), 40–42
    Jones, John Winter, correspondence with, 40–41, 42, 43
    resignation, from British Museum, 46–7
    shipments, of antiquities from Babylon, 48–52
    Telloh, excavations at, 44–5
    timeline for movements, *48*

Rawlinson, Sir Henry Creswicke, 35, 36, 36
    Birch, Samuel, correspondence with, 60, 61
    copy of Cylinder, 60
    Pinches, Theophilus, correspondence with, 60–62, 61
    presentation of Cylinder (Royal Asiatic Society), 62
Rich, Claudius James, 38
Royal Asiatic Society, presentation of Cylinder, 62
Royal Persia: A Commemoration of Cyrus the Great and his Successors (British Museum), 76, 77

Schaudig, Hanspeter, 134
Selby, William Beaumont, 38–40
    Map of the Ruins at Babylon, 57
Sennacherib, 122
Shafa, Shojaeddin, 87
Sharp, Rev. Norman, 75–6, 76, 83–4
shipments, of antiquities from Babylon, 48–52
Shuanna, 5, 6, 7, 9
Sin, 25, 105
Smith, George, 40
Smith, Sidney, 71
stamps, commemorative (Persian Empire), 90–91, 90, 91, 92, 93
Sumer, 5, 6, 7, 9
Susa, 6, 9

Tehran, loan of Cylinder
    in 1971, 76, 88
    in 2010–11, 91–2, 93–103, 95, 96, 97
Telloh, 44
Tigris, 6, 9
The Times, 35
Tintir, 5, 9
Toma, Daud, 19, 42, 45
    deliberate damage to Cylinder 55–6
    discovery of the Cylinder, 49–50
    inventory of dispatched objects, 51, 52

Verse Account, 106, 106, 107, 108, 117, 120

Wang, Jingru, 28
Wang, Nanfang, 28
The White Revolution of Iran (Pahlavi), 87
workshop, at British Museum, 2–3, 25, 94
Wu, Yuhong, 29, 30

Xenophon, *Cyropaedia*, 109, 115, 120
Xue Shenwei, 28, 29

Yale University
    Babylonian Collection, 13, 15
    Cylinder fragment, 13–15, 14
Yang Zhi, 29
Yi'an, Zhang, 28

Zamban, 6, 9